CATS VS DOGS

Also from *New Scientist*

Does Anything Eat Wasps?
Why Don't Penguins' Feet Freeze?
Do Polar Bears Get Lonely?
Why Can't Elephants Jump?
How to Make a Tornado
How to Fossilise Your Hamster
Farmer Buckley's Exploding Trousers
Chance
Question Everything
How Long is Now?
The Origin of Everything
How to Be Human
This Book Will Blow Your Mind
The Brain

CATS VS DOGS

Misbehaving mammals, intellectual insects, flatulent fish and the great pet showdown: amazing animal insights from *New Scientist*

New Scientist

First published in Great Britain in 2020 by John Murray (Publishers)
First published in the United States of America in 2021
by Nicholas Brealey Publishing
An Hachette UK company

1

From material previously published in *Does Anything Eat Wasps?*; *Why Don't Penguins' Feet Freeze?*; *Do Polar Bears Get Lonely?*; *Why Can't Elephants Jump?*; *How to Make a Tornado*; *How to Fossilise Your Hamster*; and *Farmer Buckley's Exploding Trousers*

A CIP catalogue record for this title is available from the British Library

Hardback ISBN 978-1-529-33920-8
UK eBook ISBN 978-1-529-33921-5
US eBook ISBN 978-1-529-34148-5

Typeset in Bembo by Palimpsest Book Production Ltd, Falkirk, Stirlingshire

Printed and bound in Great Britain by Clays Ltd, Elcograf S.p.A.

John Murray policy is to use papers that are natural, renewable and recyclable products and made from wood grown in sustainable forests. The logging and manufacturing processes are expected to conform to the environmental regulations of the country of origin.

John Murray (Publishers)
Carmelite House
50 Victoria Embankment
London EC4Y 0DZ
www.johnmurraypress.co.uk

Nicholas Brealey Publishing US
Hachette Book Group
Market Place Center, 53 State Street
Boston, MA 02109 USA
www.nbuspublishing.com

CONTENTS

INTRODUCTION

At *New Scientist*, we always like to say that facts matter. Since the magazine was founded in 1956 for 'all those interested in scientific discovery and its social consequences', our mission has been to bring facts, evidence and rational thinking to bear on the weightiest issues of the day, from the dawning of the nuclear age and the cold war space race, to climate change, the search for sustainable energy sources and the odd global pandemic.

And, indeed, whether cats are better than dogs.

I well remember when this first became a topic of conversation in the *New Scientist* office one lunchtime about ten years ago. We all know the world divides into cat people and dog people, and for us it was no different. Arguments were advanced, battle lines were drawn, tempers frayed. The result was an unsatisfactory stalemate.

Here, though, was surely a question where science could be usefully brought to bear. Check mission statement. And so we devised a scheme of ten objective criteria by which to judge our furry friends, and set out to contact the world's leading experts on canine and feline behaviour to settle the question once and for all.

It was perhaps, with hindsight, a mistake that the editor

in charge was vehemently a dog person. I recall her disappointment when the scores were on the kennel doors, and it was a 5–5 draw. (That was one point to the dogs, incidentally, 'tractability' – you can keep them in kennels and they won't complain.) I also remember my surprise at seeing the article in the magazine. A mysterious eleventh category had been added to gift the mutts the victory. Facts matter – clearly – only up to a point.

Far be it from me to impugn a colleague's journalistic integrity, however; all's fair in (pet) love and war. And while few questions seem to get animal lovers so hot under the collar as that of cats vs dogs, the animal kingdom more broadly is an endless source of fascination for *New Scientist* readers. We know that from how often fur and feathers feature in our weekly 'Last Word' column, in which readers send in everyday science questions to be answered by others. The 'Last Word' column has been the source of a series of bestselling books over the years, all of which tended to have animal-themed questions as their titles, many of them reproduced here. Do polar bears get lonely? Can elephants jump? Why don't penguins' feet freeze? And does anything eat wasps?

The idea of this book was to collate some of the best of those animal-themed queries in one place, together with a few bits and bobs that have popped up in the magazine over the years – catnip, we hope for the animal lover.

But the informative, intriguing and often witty insights of our readers take pride of place, providing a breadth of wisdom that's hard to achieve by conventional (or unconventional; see above) journalistic methods alone.

Or, indeed, asking the questions it never occurs to

journalists to ask. Can hamsters solve the climate crisis? Do magpies prefer semi-skimmed milk? Has a fish ever been struck by lightning? And where are all the Argentinian ants? (That one delightfully answered by Harvard entomologist Edward O. Wilson, the world's foremost authority on ants.) I never thought I needed to know those things – but I think I do now. The joy of science.

What makes animals so fascinating? (That's me asking a question now.) My guess is it's because we can never be them. The philosopher Thomas Nagel once asked another famous rhetorical question, 'What is it like to be a bat?', when grappling with the intractable problem of what consciousness is, human, animal or otherwise. The short answer is that we can't know – we are left with a fascinating guessing game about why animals behave as they do, and what they're thinking and feeling as they do it.

That is a reminder that we are just one species among many on a startlingly, beautifully, biodiverse world. It's up to us to ensure it stays that way. To strike a more serious note, much has been made of how we are entering a new era in Earth's history, the Anthropocene, in which human activities are likely to dominate the geological record. Accompanying it, unless we learn to live more in harmony with the natural world on which we ultimately depend, could be a sixth mass extinction of life – the only one caused by the activities of a dominant species. For every entertaining question about the animal world around us, there are serious questions to be asked about our treatment of it.

But let's stay on the side of delight for the moment. It only remains for me to thank all my *New Scientist* colleagues, our readers and all contributors to the magazine over many years who make books like these possible. Thanks also to Kate Craigie and Georgina Laycock at John Murray for making this one happen. I hope you enjoy it – and if you'd like your own question, big or small, answered, why not email it to lastword@newscientist.com?

Just don't ask about cats vs dogs.

Richard Webb, Executive Editor,
New Scientist 2020

HUMAN’S BEST FRIENDS

DOGS, CATS & OTHER PETS

Why do dogs like jumping into cold ponds, while cats and humans generally do not?

James Scowen
London, UK

Your questioner appears to be confusing willingness with enjoyment. Most dogs are prepared to dive into cold water, but they may not like the experience. And in referring to cats, your questioner is almost certainly referring to the domesticated species, which is not necessarily representative of its genus.

Nonetheless, the canine tolerance for cold water, and feline intolerance, lie in their respective evolutionary histories. The dog (*Canis lupus familiaris*) originated in central Asia during the aftermath of the last ice age, at least 15,000 years ago. It is descended from the grey wolf (*Canis lupus*), with all the evolutionary baggage that implies. Ice-age wolves preyed on sub-Arctic herd animals such as elk, reindeer and caribou, which would have migrated in search of better grazing, crossing fast-flowing rivers swollen by meltwater when required.

Any animal – including the ice-age wolves – fording or swimming these rivers would have had to develop considerable physical and psychological resistance to low temperatures. Those that weren't prepared to get their feet wet wouldn't have lasted long enough to pass on their genes. Those that did bequeathed their doggy descendants a tolerance for cold water.

Some 5,000 years after the big bad wolf began the transition to being man's best friend, a group of wild cats (*Felis sylvestris*) in what is now western Asia apparently attached themselves to the local human population in a semi-symbiotic relationship. Significantly, the closest living relative of the proto-kitties is believed to be the sand cat (*Felis margarita*), a denizen of regions of extreme heat and aridity, such as the Sahara.

This ancestry was never likely to cultivate a hereditary tolerance for getting wet, even if natural selection had not already instilled a wariness of bodies of water, whatever their temperature. Large mammals have no freshwater predators in the sub-Arctic, but animals originating in the tropics have good reasons for not going into the water, most of them possessing very powerful jaws. A prehistoric African water hole was a fast-food outlet for large predators, both in and around the water.

The behavioural heritage of these widely differing ancestries can be most clearly observed when our modern-day pets are drinking. A dog will generally lap up its water enthusiastically, albeit with the occasional sideways glance at any animal that could attack. A cat, on the other hand, displays far more caution, constantly looking around suspiciously and keeping its body as far back as possible from the liquid.

Hadrian Jeffs
Norwich, Norfolk, UK

Dogs, like humans and cats, exhibit a homeothermic mode of temperature regulation – their body temperature remains constant in spite of fluctuations in the temperature of their environment.

Dogs are covered with thick hair to conserve internal heat, and regulate their body temperature through panting, an extremely efficient method. On a hot day it is quite common to see a dog with its mouth wide open and tongue hanging out.

Recent research has also indicated the presence of a complex network of blood vessels in the basal part of a dog's neck. This region functions as an efficient temperature regulator. In addition, dogs have relatively large spleens. When a dog is active or under stress, the spleen contracts and releases blood into the circulatory system, which provides yet another mechanism for carrying excess heat to the skin. All this means that dogs are better adapted than humans or cats to withstand cold shocks or hypothermia.

Saikat Basu
Lethbridge, Alberta, Canada

Can I ensure my hamster becomes a fossil?

Remember, whatever applies to a hamster applies to humans too. So, if you would like to become a fossil, pay attention to what follows, make detailed notes in your will and find a sympathetic funeral director.

What do I need?

a naturally deceased hamster (or other pet)

a variety of environmental conditions

What do I do?

Take your hamster to one of the natural environments described below and ensure that it will not be removed from its final resting place by scavengers or natural phenomena.

What will I see?

Very little. Fossils take tens of thousands of years to form, but you will be saving up enjoyment for future generations of palaeontologists.

What is (should be) going on?

A desire to preserve our fluffy pets' remains in fossilised form may be admirable, but the fact that they have a hard, mineralised skeleton and live a non-marine lifestyle is a bad start. Terrestrial conditions are liable to erosion and significantly reduce the chance of fossilisation. The soft tissue surrounding the skeleton of mammals decays very quickly – it is usually preserved only if the animal dies at high altitude or in a freezing environment such as a glacial crevasse or in the polar regions, and then only in wizened, mummified form, which is not true fossilisation.

So, if you really want your pet to survive the ravages of geological time, then while it is still alive you need to concentrate on improving the quality of its teeth and bones. Fossilisation of these involves additional mineralisation, but you can give your hamster a head start by feeding it a diet rich in calcium to build up its bones and teeth. If you are seriously thinking of becoming a fossil yourself, you should get a good dentist. If your hamster has eaten a few seeds in its last days, these too can become fossilised and intrigue palaeontologists in the millennia to come.

After that it's a matter of location, location, location. You need to bury your dear departed rodent where it won't be disturbed for a very, very long time. Fossilised remains are often found in caves, so you could take up potholing to scout out suitable locations (after proper training, of course). Alternatively, you need to find a spot where your hamster will be buried quickly after it has been laid to rest – preferably somewhere natural and dramatic, the sort of site from which you can detect a distant volcanic rumble and clouds of ash being emitted skywards. Don't get too close though: an ash burial is good, incineration by flowing lava is not. Again, be very careful; you don't want to join your hamster in fossilised harmony at this stage.

As the volcanic option suggests, you may have to travel a long distance to find the optimal conditions. A desert wadi in the flash-flood season offers a good environment, as does a tropical river floodplain during heavy rain so you can bury your hamster in fine, anoxic (oxygen-free) mud.

However, the best environment to ensure a fossilised hamster is probably a sea burial in very deep water (shallow marine conditions are turbulent and full of life which will disturb or eat the remains). There are few creatures in the deep sea, and even fewer below the sediment and mud. As long as your hamster is not buried near a tectonic subduction zone where the Earth's crust is being consumed, such as the Pacific coast of the Americas, it should rest undisturbed until fossilisation takes place.

This environment and the land-bound ones described above are ideal. The oxygen-free conditions will slow decay of the body and the ash or fine seabed clay will help to preserve the body structure. Fossilisation will then proceed until your hamster is nothing more than an outline of carbon and petrified body fluids, thanks to compaction from the weight of sediment that settles above. You should allow at least 200,000 years for this.

Of course, there is more chance of winning the lottery than you or your pet ending up as a fossil, but that's no reason not to try.

P.S. The following advice to let stalactites do the work of an embalmer is found in a poem written by Richard Whatley in 1820. The subject of the poem, William Buckland (1784–1856), was one of the most famous geologists of his day – and a noted eccentric, who claimed to have eaten his way through the entire animal kingdom. His contemporary, Augustus Hare, recorded how Buckland came upon a casket containing the preserved heart of a French king. Buckland exclaimed, 'I have eaten many strange things, but I have never eaten the heart of a king before,' and promptly gobbled it up, the precious relic being lost for ever.

'Elegy Intended for Professor Buckland'

Where shall we our great Professor inter,
That in peace may rest his bones?
If we hew him a rocky sepulchre,

He'll rise and break the stones,
And examine each stratum that lies around—
For he is quite in his element underground.
If with mattock and spade his body we lay
In the common alluvial soil,
He'll start up and snatch those tools away,
Of his own geological toil;
In a stratum so young the Professor disdains
That embedded should lie his organic remains.
Then exposed to the drip of some case-hardening spring,
His carcase let stalactite cover,
And to Oxford the petrified sage let us bring,
When he is incrusted all over;
There, 'mid mammoths and crocodiles, high on a shelf,
Let him stand as a monument raised to himself.

Finally, and it should go without saying, if you intend to attempt fossilisation of your pet, wait until it has died of natural causes before carrying out the above instructions.

Why are dogs' noses black?

Rachel Colin (aged 11)
Eudlo, Queensland, Australia

While a majority of dogs have black noses, not all do. The noses of dogs such as vizslas and weimaraners match their coat colours – red and silver, respectively – and it is not unusual for puppies of any breed to start out with pink noses that then darken as the animal matures. I had a Shetland sheepdog that retained pink on the insides of her nostrils for the whole of her life.

Dogs have most likely developed black noses as a protection against sunburn. While the rest of the dog's body is protected by fur, light-coloured noses are exposed to the full force of the sun's rays. Pink-nosed dogs, hairless breeds and dogs with very thin hair on their ears need to be protected with sunscreen when they go out of doors, just as humans sometimes do, or they risk the same sorts of cancers and burns.

In addition, dog breeders have long singled out a black nose as the only acceptable colour for many breeds. Though this is based on nothing more than an aesthetic preference, it still serves as a selective influence for people breeding pedigree dogs. This adds a bit of human-directed evolution to what was already a natural tendency towards black noses.

Julia Ecklar
Trafford, Pennsylvania, USA

Black nose leather contains the skin pigment melanin, specifically in its dark brown or black eumelanin forms. Melanocytes, the cells that produce the raw material, secrete it into the skin cells, and the sun then darkens it further. Melanin in skin cells protects the DNA in cells from mutations caused by ultraviolet radiation from the sun.

Jon Richfield
Somerset West, Western Cape, South Africa

Could you breed dogs for intelligence?

Dog breeding often gets a bad press, including the apparently unfounded assertion that breeding for looks has an adverse effect on intelligence in dogs. Just how intelligent could any species get through selective breeding? And how quickly?

John Schofield
London, UK

Intensive breeding for looks in any animal adversely affects intelligence and every other attribute – eventually including those very looks. This is intrinsic to selection, whether natural or artificial.

The effectiveness of selection depends on the range of relevant genes in the population: the larger the natural population, the greater the range of genes is likely to be. Selection for any desired attribute rapidly reduces that range: in a single generation, less than one per cent of a population might be selected, immediately reducing the range of 'irrelevant' genes, including genes for mental or physical health or functionality.

Dogs bred for show are commonly selected so obsessively

that any harmful genes they carry become fixed in their populations. In competitive show breeding, selection is particularly stringent, with the result that gene pools shrink rapidly. Most mutations and recessive genes in small, closed populations are harmful, so progress is overwhelmingly negative.

The closest we come to breeding for intelligence and functionality in dogs is in certain working breeds. Breeding companion animals specifically for desirable behaviour, intelligence and health should be gratifying, but it is also challenging and commercially precarious. People who need companions prefer to buy mongrels.

Jon Richfield
Somerset West, Western Cape, South Africa

Asking if anyone has bred a variety of dog purely for intelligence begs the question of what is meant by intelligence. The psychologist Robert Sternberg has shown that what we think of as 'intelligent' depends on what we value – specifically what we think people should be good at. So, what we consider to be a clever dog would be one that does a good job at what we want it to do: herd sheep well or guard the house effectively. We have no need for dogs that are adept at calculus or playing the futures market, so we have never tested our capacity to breed this into them.

Sternberg identified three signs of intelligence: the ability to adapt to environments, the ability to shape environments and the ability to understand that the environment is not optimal, thus facilitating a move to a more congenial niche. On this basis, you could make the argument that almost all species are intelligent, even bacteria, because at the very least they are adapted to their environments. In addition, many can up sticks when things are not so good and move elsewhere, and some can even shape their environment in some way congenial to them.

Maybe dogs deserve special mention because they have shaped their environment by making themselves useful and appealing to humans, in return for food and shelter.

Catherine Scott
Surrey Hills, Victoria, Australia

Cats all seem to like fish, so why are they unwilling to swim?

Tom Lorkin
Beckenham, Kent, UK

I like fish too, but you won't catch me on a trawler, let alone in the water. For one thing, like most cats, I simply cannot swim well enough to catch any meal that swims away from me. While otters, seals and other aquatic mammals and reptiles swim after the fish they hunt, there are comparatively few of them. Some cats fish actively, but they do so from the bank, or leap on fish in shallow water. Whether cats in a particular region catch fish depends on their having learned the skill. Where fishing is good, cats of all sizes from wildcats to jaguars may snatch fish, and the south-east Asian fishing cat apparently does so frequently.

As for being unwilling to swim, cats vary. Some actually like it. I have seen the hilarious sight of a moggie lowering itself into a swimming pool tail-first to avoid getting a nose full of water, swim around to cool off, then head for the steps. The breed commonly called the Turkish Van is well known for enjoying the occasional summer dip.

Jan Rhode
Exeter, Devon, UK

Cats are quite capable of swimming if they have to but may dislike it because of its effects on their fur. A cat's fur is effective insulation both from the cold and the heat, thanks to the way it lies on the cat. If a cat gets soaked, the fur becomes waterlogged and the cat can lose body heat to the extent that it becomes hypothermic. However, while a cat will seek shelter in the rain, a little damp does no harm because the top layer of fur is water-repellent and rain just bounces off. For this reason, it's not a good idea to dry a moderately wet cat with a towel because water will get through the water-repellent layer to the more absorbent hairs below. If a cat is really wet, it's best to dry it with a hairdryer on a very low setting. However, most cats are frightened of this, so letting it sit in front of a fire is probably better.

Cats can also fish. About twenty years ago my family had a cat that regularly brought back bullhead fish. Cats sit on the riverbank and when a fish comes into range the cat hoicks it out with extended claws and throws it over its head and clear of the water. The fish is then helpless and the cat has its meal or trophy.

Charles Stuart
Bath, Somerset, UK

Snow leopards, lynx and other species from cold environments avoid getting wet because water compromises the ability of their fur coats to keep them warm. On the other hand, lions, tigers, jaguars and other species that live in hot habitats often take a dip to cool off. It is thought that the Turkish Van, which hails from the region around Lake Van in eastern Turkey, took to swimming to escape the scorching heat. This swimming cat has dispensed with the undercoat that most cats have and its fur has a cashmere-like texture that makes it water-resistant. The fishing cat (*Prionailurus viverrinus*) from south-east Asia has gone one step further and dives into water to catch fish.

Fishing cats have been reported to attack ducks from under the water.

While a Turkish Van can go for a dip and come out relatively dry, most domestic cats hate getting wet, possibly because they must spend hours putting their fur back in order. However, some domestic cats will happily join their owners in the shower or play with a dripping tap.

Mike Follows
Willenhall, West Midlands, UK

On one occasion my cat swam out to my fishing boat, a distance of about 100 metres, presumably for company and a feed of pilchards. Her swimming style is similar to a doggy-paddle. She only breached the surface when taking a breath (in a similar manner to a seal). On another occasion, we were netting for bait, and she swam behind the net, attacking fish that were caught in it.

I guess some cats swim and some cats don't like fish.

Richard
By email, no address supplied

Should I feed my dog tomato ketchup?

I have two female dogs whose urine kills the grass in patches all over the lawn. My mother advised feeding them tomato ketchup, which I did, and the patches stopped appearing. Why does this work, and should I really be feeding my dogs tomato ketchup?

Jim Landon
Swindon, Wiltshire, UK

The urine acts as a liquid fertiliser, but can produce nitrogen overload where the puddle of urine is deepest. This 'burns' the grass, creating a brown patch in the lawn.

Towards the outside of the puddle, where less nitrogen has been applied, there can be a fertilising effect leading to a ring of luxuriant, greener grass. The urine of dogs and bitches does not differ much but, while dogs tend to deliver small samples of urine to mark their territory, bitches tend to empty their bladders entirely, causing more harm.

Urine is slightly acidic, but so is tomato ketchup, so it does not neutralise the urine as some people believe. Instead, the salt content of tomato ketchup, juice or sauce makes dogs drink more, diluting the nitrogen in their urine.

Be aware that increased salt intake can cause problems with existing kidney or heart conditions, so if you must tinker with your dogs' diet, consider reducing the protein content instead. This will also reduce the nitrogen content of their urine, and should be fine for all but the most active of dogs. Better still would be to train your dogs to urinate in a designated place or follow them out of the house with a hose pipe or watering can to dilute their urine.

Mike Follows
Willenhall, West Midlands, UK

Are hamsters the answer to solving the climate change crisis?

Could hamster power be an environmentally friendly answer to the impending energy crisis? How many hamsters running on wheels would it take to provide energy for a house or a factory?

Catherine Hetherington
Aberdeen, UK

Let's assume a hamster weighing fifty grams can run up a thirty-degree slope at two metres per second. This corresponds to a power output of half a watt. If it delivers the same power when running in a hamster wheel, we would need 120 hamsters working flat-out to light a sixty-watt bulb.

The average hamster probably doesn't spend more than 5 per cent of its life running in its wheel, so already we need a brigade of 2,400 hamsters just to light our bulb. It gets worse. The average UK household consumes in excess of eighty gigajoules of energy per year. This is equivalent to a constant power consumption of about 2.5 kilowatts. Each house would need 100,000 hamsters. Multiply this by the number of households in the UK and we would have an environmental and economic disaster.

In addition, we would need to employ an army of animal behaviourists to devise Pavlovian tricks to get the hamsters onto their wheels in response to surges in demand. And given that hamsters are nocturnal, this would force politicians and lawyers to debate animal welfare. The UK alone would need to employ everyone else in Europe to feed and care for its hamster population.

Perhaps we should let humans run on treadmills. It would not produce much electricity but we might end up with less of an obesity problem.

Mike Follows

Willenhall, West Midlands, UK

Hamsters running on wheels cannot relieve the impending energy crisis because animals are not energy sources – they are energy consumers. It would be more efficient to simply burn their food in a furnace and use the power output from that (a fact overlooked in *The Matrix* films, where humans are used as thermal energy sources by sentient computers).

John Woods
Stratford-upon-Avon, Warwickshire, UK

According to the CIA website, the estimated global electrical energy consumption in 2003 was 15.45 trillion kilowatt-hours. To produce that kind of energy in ideal conditions would require around 1,458 billion hamsters. Hamsters have an average lifespan of 2.5 years, meaning that if we had switched to hamster power in 2003, we would already have more than two billion tonnes of depleted hamster, and many backyard funerals. The environmental and socioeconomic impact of this would be devastating. So it is my duty as a pseudo-technician to decree that this is another energy source best left to fiction.

Ben Padman
Perth, Western Australia

The question is not whether hamster power is an environmentally friendly energy source, but whether it would be welfare friendly. As a veterinary student, I have spent some time looking at research into 'the running wheel phenomenon'. It is clear that captive hamsters are highly motivated to use running wheels. What mystifies researchers is why.

There is controversy over whether running-wheel activity is a stereotypic behaviour – repetitive, invariant behaviour with no obvious function (likened by some to obsessive-compulsive disorder in humans) which results from a sub-optimal environment. Even if running-wheel use turns out not to be a stereotypy, there is further debate as to whether it corresponds to poor welfare, because such behaviours may merely be a way of coping with captivity.

However, what has been found is that when hamsters are given bedding that is eighty centimetres or more deep, which lets them indulge in natural burrowing behaviour, their use of running wheels drops dramatically and the performance of other stereotypies such as wire-gnawing ceases altogether. This suggests that we should reconsider how pet hamsters are kept. Perhaps finding a way to harness the burrowing activities of hamsters would be a better solution to the energy crisis.

Sarah Briars
Shefford, Bedfordshire, UK

Total world annual energy consumption is about 500 exajoules (an exajoule is 1,018 joules). A hamster requires fifteen grams of food per day. Let's assume that the hamsters eat wheat, with an energy content of 1400 kilojoules per 100 grams. If we assume that they convert the chemical energy of their food into useful energy with the same efficiency that power stations and wood-burning stoves do, some 6500 billion hamsters on wheels would be needed to supply the world's energy requirements. On a more manageable mental scale, energy use in a typical house in the UK is about eighty gigajoules a year, which is the amount 1000 continuously running hamsters would produce.

The drawback to maintaining 6500 billion hamsters is that worldwide they would need 36 billion tonnes of wheat per year, nearly sixty times the world's present wheat production.

This exercise illustrates that the world's energy crisis is not simply due to excessive use of fossil fuels, to be solved by conversion to renewable energy sources. The scale of such a conversion is far too great, and large-scale renewables have their downsides too. Cutting energy use through conservation and, by implication, a change in the way we live, is the main answer to the energy and climate change crisis. This is why politicians find it nearly impossible to confront the issue.

Philip Ward
Sheffield, South Yorkshire, UK

Is there ash in dog food?

Almost all dog food contains ash as an ingredient. Sometimes it makes up as much as fourteen per cent by weight. I have always thought of ash as being toxic waste, containing all sorts of noxious elements, so why is it added to dog food and what type of ash is it?

William Davidson
Strathaven, Lanarkshire, UK

You will be relieved to hear that ash is not added to pet foods. It is a way of describing the mineral content of pet food. The ash you see listed is part of the guaranteed nutrient analysis: legally the pack must state how much of the food is protein, fat, fibre, water and ash.

Ash is measured by heating the pet food to temperatures of around 550°C, and burning off all the organic components to leave just the inorganic or mineral residue. If the mineral content of pet food sounds high, it is important to remember that our domestic carnivores were designed to eat carcasses that are full of bones containing minerals, and a well-designed pet food will reflect this in its composition.

Kim Russell
Registered pet nutritionist
North Molton, Devon, UK

This is a misreading of the label on the product. Ash is usually given under 'typical analysis' or a similar heading, not under the ingredients list.

Foods are often described in terms of their nutritional content by carrying out a proximate analysis. This is done because it is much quicker and cheaper than carrying out a detailed analysis of the nutrients.

The protein content is determined by analysing the nitrogen content, using a technique called the Kjeldahl method, and multiplying by a conversion factor to obtain a crude protein figure.

The fat content is derived by gravimetric extraction using a suitable non-polar solvent (usually petroleum ether), while the water content is obtained by drying, and the mineral content is found by burning off all the organic material in a muffle furnace to obtain the ash. The carbohydrate content is often estimated by subtracting the former components from the total weight.

So ash is not added as an ingredient but is instead an indicator of mineral content. These minerals will be chiefly potassium and phosphorus with smaller amounts of calcium, iron, magnesium, sodium and zinc, and trace amounts of many others. Historically, manufacturers often boosted the mineral content of dog food with bone meal to raise calcium levels but, because of concerns about BSE, they now tend to use fish meal instead.

You might see ash levels of 14 per cent in a dry meal for dogs, but tinned products often have around half this level. The composition of the food will affect the ash content, but the elements are likely to be beneficial or neutral to the dog's health and not noxious or toxic at the concentrations in the product. It is worth pointing out that dogs should always have access to fresh water to ensure they can urinate away any excess potassium or sodium.

Brian Ratcliffe
Professor of Human Nutrition
The Robert Gordon University
Aberdeen, UK

How long can you keep a tiger cub as a pet?

I have read of people doing so, but surely, for very obvious reasons, there is a time limit to how long you can keep a carnivore in your living room.

Peter Higginson
Paris, France

If there is an upper age limit then one is assuming that tigers are suitable for a domestic environment below that age. The flaw with this is that they would need their mother at this stage and, if one has qualms about keeping an adult tiger, then one would not even contemplate a maternal tigress.

As for the implication that there is a stage in a juvenile tiger 's physical development which would mark a watershed in pet–owner relations, then it is a case of 'take your pick'. Cubs become fully mobile at about eight weeks, when they are still endowed with cuteness, but you wouldn't want one to bite or scratch you. At eighteen months, young tigers become independent in terms of being capable of fending for themselves. Fending for themselves in this context generally means hunting. That does not mean they have left home, however. In the wild, cubs stay with mothers for up to two and a half years, and removing one may have emotional repercussions for an animal that is already a dangerous predator.

Even if you and your family avoid featuring on the tiger's menu, there are the minor social, and, probably, legal difficulties arising out of the disappearance of your neighbours' pets. Tigers do not only kill to eat; in the wild they will also suppress local populations of any rival carnivores in their territory. These include wolves – and next-door's labrador. And, if it's a male tiger, there's the particularly noxious spray-marking of personal space to consider – tiger pee makes fox urine smell like Chanel No. 5.

Clearly it is not impossible to keep a tiger as a pet. According to the most recent statistics from the US Association of Zoos and Aquariums, some 12,000 are kept as pets in the USA alone, ironically more than the world population of wild tigers, and largely as a consequence of overbreeding by zoos in the 1980s and 1990s. How many are kept as domestic pets in the accepted sense, rather than in wildlife parks or private enclosures, isn't known.

However, in a sad commentary on US animal welfare legislation, a large number of these big cats must be living in people's homes, as if they were large dogs. For instance, the American Society for the Prevention of Cruelty to Animals estimates there are 500 lions and tigers in metropolitan Houston alone. While nineteen states have banned private ownership of big cats, only fifteen of the others require an owner's licence, and sixteen have no relevant legislation at all, despite the USA's indigenous population of cougars being on the endangered species list.

These animals are all adults, so strictly speaking there is no upper age limit for keeping a tiger as a pet because it has grown too large, or its behaviour makes it unsuitable. The correct answer to the question is, of course, that no tiger of any age should live socially alongside people. Their proper place is in the wild.

Hadrian Jeffs
Norwich, Norfolk, UK

You can keep a tiger cub as a pet until it grows up and gets hungry or loses patience with you.

Doug Grigg
Cannonvale, Queensland, Australia

Does a dog know it's a dog?

My four-year-old daughter asked me if her dog knows that it is a dog. Does her pet realise it is different from us or does it think that we're just odd-shaped dogs or, indeed, that it is a particularly impressive human being?

Celia Denton
Stillington, North Yorkshire, UK

The question calls to mind cyberneticist Stafford Beer, writing in a 1970s edition of *New Scientist*: 'Man: "Hello, my boy. And what is your dog's name?" Boy: "I don't know. But we call him Rover."'

The boy's reply reveals his belief that his dog has a mental image of itself (which he assumes to include a name), but at the same time confesses his inability to penetrate the dog's psyche. And he's right: the short but unsatisfying answer to the question above is that we don't know what goes on in a dog's mind.

We can, though, make a reasonable stab at it. For a start, it's clear that when individuals of different species interact, judgements of sameness or difference are simply not part of the story. That the tiny reed warbler heroically feeds the gigantic cuckoo nestling, despite the obvious (to humans) fact that it cannot possibly be a warbler, indicates that the bird isn't operating according to any concept of warbler-ness or cuckooness, but purely to one of this-thing-needs-to-be-fed-ness.

Dogs have specific responses to things-that-dogs-can-eat (such as rabbits) and things-that-can-eat-dogs (such as lions), and also to potential mates or rivals, and to offspring. Other than that, they resemble humans in viewing a variety of creatures of whatever sort as potential social companions or friends. Indeed, that is why you have a dog – and the dog tolerates you – in the first place. As with humans, the establishment and maintenance of social bonds is key to dogs' way of life. I think your dog sees the questioner as her friend in much the same way that the questioner sees the dog as hers. Lucky dog, lucky you.

Angus Martin
Camberwell, Victoria, Australia

Canids in general start to develop social relationships when their eyes and ears open at about two weeks of age. During the critical period between two and sixteen weeks, puppies learn the social rules that will shape their behaviour for the rest of their lives, including recognition of conspecifics and appropriate mates. The famous ethologist Konrad

Lorenz, when studying greylag geese, found that sexually mature geese raised by a human 'mother' tended to direct their courtship behaviour toward humans rather than other geese.

In dogs, this same confusion can be seen in the way dogs direct social dominance and play behaviours toward humans – in effect, treating people as if they were dogs. Likewise, livestock-guarding dogs, such as those protecting sheep, are trained for their jobs by removing them from their mothers at just a few weeks of age and allowing them to grow up with sheep as their companions. The sheep are then forever recognised as family and are socialised with and protected as such.

After 16 weeks, this period of rapid learning and adaptation ends, and the social skills the dog has are pretty much set for life. This is why it is so important for puppies to have intimate contact with people from the time they are born. Traditionally, we adopt pet dogs when they are eight or nine weeks old, right in the middle of this period of social development, and proceed to lavish them with attention and experiences through to the end of that sixteen-week period. The result is that the dog in the question above sees nothing at all odd about her tall, hairless pack-mates.

Julia Ecklar
Trafford, Pennsylvania, USA

Can humans stay cool by panting like dogs?

In hot weather dogs keep cool by panting. If I were to do this I would hyperventilate and exhale too much carbon dioxide. How do dogs avoid the effects of respiratory alkalosis?

Andrew Benton
Birkenhead, Merseyside, UK

Each breath taken by a human (or a dog for that matter) consists of a volume of air that enters the lungs and a smaller volume that only gets as far as the passages that lead to them. This is the 'dead space', so called because no exchange of oxygen and carbon dioxide occurs in the mouth, pharynx, trachea or bronchi.

Rapid, shallow breathing can affect just this dead space without hyperventilating the gas-exchange part of the lungs, the alveoli. As air passes through the dead space it produces a cooling effect as moisture lining these passageways evaporates. Dogs, lacking sweat glands, use this method to cool down. Humans have no need of this because we use sweat to cool our bodies, though we can do it. Try 'fluttering' your breathing by taking fast, shallow breaths, at least sixty per minute. You will feel a cooling effect in your mouth, but not the dizziness that can accompany hyperventilation. It's hard work though . . .

John Davie
Anaesthetist
Lancaster, Lancashire, UK

Dogs vs cats: The great pet showdown

The world is divided into 'dog people' and 'cat people', each passionately believing that their preferred pet is superior. Until a decade ago, there was very little scientific evidence either camp could muster to support its claims. Then animal behaviourists became interested in dogs and unleashed a pack of ingenious experiments testing canine capabilities and cognition. Recently, researchers have started doing similar work with cats. Could it be time for that showdown?

There are obvious pitfalls in trying to use science to resolve this perennial dispute. Every pet owner knows their furry family member is special – a unique being with its own talents and foibles. Yet scientific research tends to look at species as a whole and deals in averages and trends when attempting to quantify their characteristics. Then there is the thorny issue of comparing two very different animals. Some might argue that the whole venture is doomed to failure, but here at *New Scientist* we like a challenge. So we have pitted cats against dogs in eleven categories. It's a winner-take-all competition with 'best in show' being awarded to the pet that prevails in the most categories. Let the fur fly . . .

1. BRAINS

At sixty-four grams, the average dog brain is far bigger than its feline equivalent, which weighs in at a mere twenty-five grams. But then the average dog is much heavier than the average cat. If instead you measure brain mass as a percentage of body mass, cats win by a whisker.

Felophiles should not gloat yet. In general, smaller mammals have slightly larger brains relative to their body size than bigger ones. This means cats' brains are exactly the mass you would expect for their size, whereas dogs have slightly more upstairs than you would predict.

On size alone, then, the results are ambiguous. That is perhaps all to the good, because brain size is not a reliable measure of intelligence. In fact, if you want to assess smarts you are far better looking at behaviour than crude neuro-anatomy – more on that later. However, there is one anatomical measure that gives a pretty good indication of information processing capacity: the number of neurons in the cortex, or executive brain. Here cats trounce dogs, with 300 million neurons compared with a piddling 160 million.

WINNER: CATS

RUNNING SCORE: CATS 1 – 0 DOGS

2. SHARED HISTORY

Several research teams have compared DNA from dogs with that of grey wolves, their closest living ancestor, to try to pinpoint the date of domestication.

In the first study of this kind in 1997, Robert Wayne's team at the University of California, Los Angeles, came up with a date of 135,000 years ago. Since then, the entire dog genome has been sequenced and Wayne now believes his date may be a little premature. Nevertheless, given the discovery of archaeological remains of dogs dating from 31,000 years ago and the large divergence between dog and wolf DNA, he still suspects that domestication occurred at least 50,000 years ago.

Other DNA studies, however, suggest domestication could be more recent. A recent analysis, published by Peter Savolainen at the Royal Institute of Technology in Stockholm, Sweden, comes in at 16,000 years. It also points to an origin south of the Yangtze river in China and speculates that the first dogs were not working dogs, but destined for the dinner table.

Our knowledge of feline domestication is also fuzzy. Evidence from ancient Egyptian burials and hieroglyphs indicates that cats were popular in homes from about 3000 BC onwards. However, the synergy with humans probably stretches further back. As soon as our ancestors began farming, their grain stores would have become magnets for vermin, and therefore cats. In 2007, comparisons of the DNA of wildcats from across the globe with that of domestic cats confirmed their origins in the Fertile Crescent east of the Mediterranean, the cradle of agriculture. What's more, cats seem to have wormed their way into our homes and hearts from an early stage, as evidenced by a 9500-year-old burial of a kitten alongside a human on Cyprus. While impressive, it still leaves Felix looking like a newcomer compared with Fido.

WINNER: DOGS
SCORE: CATS 1 – 1 DOGS

3. BONDING

The bond between a dog and its owner is remarkably similar to that between a parent and child. A secure baby behaves in a characteristic way in strange situations: it is courageous and happy to explore while its mother is around, becomes distressed when she leaves, will settle with a stranger in time, but has eyes only for mum when she returns.

Dogs put through the 'strange situation' test respond in the same way. That is probably no surprise to dog people, who often cite unconditional love as their pet's more endearing quality. Are cats so very different?

Ádám Miklósi from Eötvös Loránd University in Budapest, Hungary, whose group did the work with dogs, tried this experiment with cats – but they were having none of it. The lab setting was very upsetting and stressful for them, presumably because cats tend not to leave their territory. Nevertheless, Miklósi suspects that cats bond with their owners in much the same way that dogs do – if only he could persuade them to take the test.

Even the most besotted owner will admit that cats like their independence. Evolution is to blame. By nature, cats are loners. Dogs, meanwhile, are descended from pack animals and have an instinct to affiliate, and domestication has changed the focus of this instinct. Give a four-month-old puppy the choice and it will choose a human companion over a dog. It seems they just can't help but love us.

WINNER: DOGS
SCORE: CATS 1 – 2 DOGS

4. POPULARITY

Arguably the ultimate test of whether an animal makes a good pet is how many people actually own them. Here cats are clear winners. Although worldwide figures are hard to come by, recent studies show that in the top ten cat-owning countries there are almost 204 million felines. Pet pooches in the top ten dog-owning countries number fewer than 173 million.

WINNER: CATS
SCORE: CATS 2 – 2 DOGS

5. UNDERSTANDING

Rico the border collie is famously able to understand over 200 words. He's a clever boy, but even dogs with more limited comprehension can often recognise and respond to dozens of commands and requests for objects. And words are not the only channel of communication open to them.

Pooches can follow human pointing gestures, such as an outstretched finger or a nod of the head, to find food. That may not seem impressive, but chimps struggle to do it. Dogs also hold eye contact with humans – which wolves tend not to do – and use gaze alternation to bring objects to their owner's attention. They seem predisposed to inspect our faces for information, reassurance and guidance, according to Alexandra Horowitz, who studies animal cognition at Columbia University in New York, and whose book *Inside of a Dog* was published in September 2009.

However, Horowitz provides a cautionary tale for anyone tempted to overestimate their dog's level of comprehension.

Her experiments revealed that a pooch's characteristic 'guilty look' does not in fact signify an understanding of transgression, but is often simply produced in response to a scolding, regardless of whether or not it has been disobedient.

Cognitively speaking, cats are similar to dogs, says Miklósi, so you would expect them to have similar patterns of behaviour and abilities. A big difference is that they are not compliant or motivated, making them devilishly hard to work with. Nevertheless, Miklósi's team found that Felix is just as capable as Fido when it comes to following pointing gestures to find food. However, if the food is hidden and impossible to retrieve, dogs are far more likely to solicit help from their owners by gaze alternation, whereas cats mostly try in vain to obtain the reward for themselves. Understanding that humans can get you what you want may seem like cheating, but add to this the dog's superior vocabulary and eagerness to engage with its owner and it is only fair they win this one.

WINNER: DOGS
SCORE: CATS 2 – 3 DOGS

6. PROBLEM SOLVING

'Cats don't understand string theory' ran the headlines in 2009 after an experiment showed that if you offer kitty a choice between two pieces of string, one with a morsel of food at the end, they often fail to pull on the string attached to the reward. Before canine-lovers crow . . . dogs do not pass the test either.

What's more, neither pet can use figurative cues to find hidden food. In other words, they don't understand X marks the spot. On the positive side, both are pretty good at retrieving bits of food from stashes placed at various distances from each other and from themselves. Although they employ slightly different strategies, their choices about the order in which they visit sites is efficient and logical.

Not much else is known of cats' problem-solving capabilities. Dogs have been subjected to far more testing, and have often failed to shine. In fact, there is a long-standing view that dogs are dunces compared with their wolf cousins, whose brains are a third bigger. One classic study showed that wolves learned to open a door with a complicated catch simply by watching another wolf do it, whereas dogs failed to master the catch even after years of seeing the door opened and closed.

But Miklósi, along with József Topál of the Hungarian Academy of Sciences in Budapest, suspected that a dog's partnership with its owner might be making it appear more stupid than it really is. The more intimate the bond, they reasoned, the more likely a dog is to relinquish its powers of independent thought and action to its owner.

Their suspicions were confirmed when they tested a variety of dogs on a task in which they had to pull on the handles of a plastic dish protruding from underneath a wire fence to retrieve some food. The most highly bonded dogs performed worst – but their success rate improved as soon as their owners encouraged them. The researchers conclude that dogs are not poor problem solvers, per se, but tend to favour a collaborative approach.

The full genius of this strategy is only revealed when you consider seeing-eye dogs. In their collaborations with blind owners, they often take the usual canine role of junior partner, but when the need arises they step in to solve problems their human cannot master. Chalk one more up to the small-brained wolf.

WINNER: DOGS
SCORE: CATS 2 – 4 DOGS

7. VOCALISATION

Shared ancestry means that all mammals tend to produce the same kinds of vocalisations to convey certain meanings. For example, they make sudden sounds with rising or rapidly fluctuating pitch to attract attention or demonstrate arousal, motivation or readiness. Both cats and dogs play on this mammalian mutual understanding in their vocal interactions with humans. Analysis of cat miaows reveals that they contain acoustic patterns that grab our attention. But the vocal repertoire of cats is quite limited and their calls tend to be idiosyncratic, so they are often interpretable only by their owners.

Dogs have far more vocal flexibility. They can vary the length, range, pitch, frequency modulation and tonality of their barks and they use this ability to produce characteristic barks in different situations. Even someone who has never owned a dog can make a good stab at telling, simply from its barks, whether it is lonely, aggressive or happy. Miklósi's group, who made this discovery, point out that other adult canids, including wolves, rarely bark. He suggests that during the course of domestication dogs may have evolved their elaborate vocal repertoire especially to communicate with us.

That's clever, but complexity is not everything. After all, no matter how much you love your pet, the barking or miaowing can get on your nerves. It looks as though cats may have found a way around this, though. A study published in 2009 reveals the subtlety with which they can use their crooning to ensnare us. By embedding an urgent high-frequency miaow into a blissed-out purr, they produce a sound that brings out our nurturing side. Karen McComb from the University of Sussex in Falmer, UK, who analysed these 'solicitation purrs', suggests they work on a subliminal level in much the same way as a baby's cry, which has a similar frequency range. For their guile, cats get the cream.

WINNER: CATS

SCORE: CATS 3 – 4 DOGS

8. TRACTABILITY

Dogs are easy to train because we have selected them to be so. They have evolved to fit into our homes and meet our needs, and they find it easy to learn and obey our rules. They are especially skilled at cognitive tasks that require cooperation and sharing information to achieve a goal.

While other animals such as chimps and dolphins learn by emulation – watching another individual carry out a task and then trying to achieve the same result – dogs learn in the same way as human infants. This process, called pedagogy, entails implicit teaching, with the dog attending to cues such as eye contact, gesture and vocalisation, and then directly imitating the actions of its master.

The most basic way to train a dog involves reinforcing the behaviours we want to encourage by giving Fido a titbit of food. Cats can be taught using rewards too. 'They respond to stimulus and reinforcement,' says Miklósi. But since no one has really tried training cats, we do not know the full extent of their abilities. Although there may be fewer ways to do it, they can probably achieve similar ends to dogs, Miklósi believes. 'But dogs really want to do it. They are more interested and take it more seriously'.

Besides, even without explicit instructions dogs naturally pick up the rules of domestic behaviour. This happens through play, according to renowned animal behaviourist Marc Bekoff of the University of Colorado, Boulder. He argues that the function of rough-and-tumble play is to develop a rudimentary sense of morality, and that such interactions with their owners allow dogs to test the limits of what is acceptable in a domestic setting. Dogs win paws down.

WINNER: DOGS
SCORE: CATS 3 – 5 DOGS

9. SUPERSENSES

Smell, sight and hearing are the most important senses for both cats and dogs. Having created endless breeds of dog to capitalise on their various perceptual talents, we should expect them to outperform the less highly selected cat – and they do show some quite amazing abilities. A bloodhound's nose, for example, contains 300 million smell receptor sites compared with just 5 million in humans. Its sense of smell is up to 100 million times more sensitive than yours.

However, while a dog's keen nose is legendary, cats are no mean sniffers either. In fact, because there is so much variability among breeds of dog, the average cat, with its

200 million smell receptors, actually has a more acute nose than the average dog.

Neither Felix nor Fido can match us when it comes to visual acuity, but their ancestry as nocturnal hunters has left them with some impressive visual abilities. Both have a faster 'flicker-fusion rate' than we do, meaning the cells in their retinas take more snapshots of the world per second than ours, giving them superior sensitivity to movement. The main reason for this is that their eyes contain many more rod cells than cones, which also explains their poor colour vision. On the upside, rods are particularly good for seeing in low light. Here, once again, cats have the upper hand. Felix can see in light levels six times as low as we can, while Fido's limit is thought to be about five times ours.

Add hearing to the list, and cats score a hat trick. Their auditory range extends from forty-five to 64,000 hertz, far wider than that of dogs at between sixty-seven and 45,000 hertz.

WINNER: CATS
SCORE: CATS 4 – 5 DOGS

10. ECO-FRIENDLINESS

Cats love wildlife – in the UK alone they kill more than 188 million wild animals each year. But dogs are no bunny huggers. They have been implicated in the decline of the rare European nightjar, they disturb ground-nesting birds and, even when walked on a lead, their mere presence may reduce biodiversity.

The real difference in ecological impact comes in consumption. A medium-size dog's ecological footprint – the area of land required to keep it fed – is 0.84 hectares annually. You could run two SUVs on that and still have change. Even a toy dog such as a chihuahua has a footprint of 0.28 hectares per year. Meanwhile, your average cat's ecological pawprint, at just 0.15 hectares, looks positively virtuous.

WINNER: CATS

SCORE: CATS 5 – 5 DOGS

11. UTILITY

Dogs can hunt, herd and guard. They can sniff out drugs and bombs and even whale faeces; they guide blind and deaf people, race for sport, pull sleds, find someone buried by an avalanche, help children learn and possibly even predict earthquakes. Cats are good if you have an infestation of rodents.

Perhaps that assessment is unfair, though. After all, we love our pets for other reasons. Cats are beautiful and soft, and stroking them has been shown to reduce stress. Then again, dogs are also good stress-busters: owning one can lower your blood pressure and cholesterol levels. What's more, Fido has other health benefits. Daily dog walks may be a chore, but they repay the effort, not just in terms of regular exercise, but also by providing immune-boosting opportunities for social contact with other dog walkers. That's why in a head-to-head contest of health benefits, it's dogs all the way.

WINNER: DOGS

FINAL SCORE: CATS 5 – 6 DOGS

Creature Curiosities

The cat's whiskers

Many inventors claim that their latest idea is 'the dog's bollocks'. But in 1996, CTI Corporation of Buckner, Missouri could claim that theirs really was. The company had developed a line in polypropylene canine testicles.

Artificial testicles, according to *Chemical and Engineering News*, were selling like hot cakes in the USA and Canada. Trade-named Neuticles, they were installed in the dog's scrotum in a two-minute procedure immediately following removal of the original articles. The idea was to be able to neuter dogs without causing psychological trauma.

Gregg Miller, CTI's president, suggested that Neuticles would encourage owners to have their dogs seen to: 'With these, the dog looks the same. He feels the same. He doesn't even know he's been neutered.'

Although Miller admitted that some people thought the product was silly, he claimed that Neuticles were 'big news in the veterinary industry'. He had even produced 'I love Neuticles' bumper stickers for the proud owners of dogs with ersatz balls. Veterinarians, however, were sceptical of the idea that Neuticles could do much for a dog's self-esteem.

Gerbils crack down

In 1982, drug smugglers at Canadian prisons and airports encountered a new force in the campaign to protect public morals – a team of highly trained gerbils. The sensitive noses of these furry detectives won £20,000 worth of government investment to help track down the nation's criminals.

A number of problems inherent in gathering a workforce of conventional dope-sniffers (dogs, for example) led the Canadians to make the breakthrough for rodent liberation. Dogs obey only one handler (or two at most); they eat a lot; and they need a fair amount of space and care. Add to that the discomfort they bring by sniffing around the ankles of harassed travellers in airports and there seemed to be a good case for the gerbil as a seeker-out of contraband.

The customs authorities set up an intensive training scheme to make the most of the gerbils' talents. At airports, where a few pioneering gerbils were in action, the sniffers crouched in their cages behind a counter and caught the smells of travellers as they walked past a fan. Although the customs gerbils faced retirement sooner than a dog would have done, officials were enthusiastic about their cheaper upkeep and their modest approach to industrial relations.

Humming dog

You would always know when Zoe the West Highland terrier was around. Zoe had two problems in life. One was that, despite the name, Zoe was male. The other was that a persistent humming noise emanated from his head – a humming noise that was audible to Zoe's owners and any other humans who happen to be around.

In 1996, his mystified owners took Zoe to a vet, Ian Millar of Belfast. After various tests and a course of treatment with antibiotics in case an infection was responsible,

Millar professed himself mystified as well. He wrote to *Veterinary Record* asking if anyone could shed light on the problem.

A couple of weeks later, Patrick Burke of the University of Edinburgh's Royal School of Veterinary Studies wrote back. The problem, he suggested, was a phenomenon known as 'otoacoustic emission'. In this condition, the normal hearing pathways in the ears are somehow reversed, so that the cochlear efferent nerve fibres stimulate outer hair cells to vibrate and make a noise. Other parts of the ear, such as the tympanic membrane, can then amplify the sound, until you end up like Zoe, humming wherever you go.

CREEPY CRAWLIES

THE MINUSCULE WORLD OF INSECTS AND ARACHNIDS

Do ants get scared?

If I saw a huge dinosaur, I would probably run for my life. So why do ants seem oblivious to a human towering over them?

Robert Watson
Jesmond, New South Wales, Australia

I think it's all relative. I measured the height of several local ants and found the largest black carpenter ants reach an average of 5 millimetres above the ground. The tallest dinosaurs were about 8 metres high. Humans are mostly less than 2 metres tall. This would make the largest dinosaur about four times the height of an average human while the average human would be about 1,000 times the size of an ant.

Some ants do, in fact, seem to sense us – especially if our shadows fall over them – and run. But mostly I suspect we are just too big to enter their awareness. Also, I wonder if ants ever look up.

Earle McNeil
Olympia, Washington, USA

Ants' attitude to life is vastly different to that of mammals, which invest a great deal of time and energy in their young and have evolved numerous means of self-protection. Ants, on the other hand, invest very little in their myriad workers, all being easily replaceable clones, so they have no individual fear of being killed. If the nest is threatened, however, it's a different story, and they will defend it to the death.

Also, ants have been around for more than 100 million years. If they think about it at all, which is unlikely, no doubt ants would see humans as a very transient species, occupying but a moment in time on their planet.

Tony Holkham
Boncath, Pembrokeshire, UK

Is the flight of the bumblebee really impossible?

My girlfriend tells me it is impossible to explain how the bumblebee flies. Apparently it defies the laws of physics. Is this true?

Torbjørn Solbakken
Norway

The infamous case of the flightless bumblebee is a classic example of carelessness with approximations. It stems from someone trying to apply a basic equation from aeronautics to the flight of the bee. The equation relates the thrust required for an object to fly to its mass and the surface area of its wings.

In the case of the bee, this gives an extremely high value – a rate of work impossible for such a small animal. So, the equation apparently 'proves' bees cannot fly.

However, the equation assumes stationary rather than flapping wings, making its use in this case misleading. Of course, if equations fail in physics there is always empirical observation – if a bee looks as if it is flying, it most probably is.

Simon Scarle
London, UK

Do beetles have a lethal design fault?

While working in the garden, I saw a beetle walk past, take a wrong step and land on its back. Without my intervention it would have stayed in this position and probably died. Why is it that millions of years of evolution have not eradicated this basic and potentially lethal design fault?

Greg Parker
Brockenhurst, Hampshire, UK

If your correspondent had left the beetle in place on its back it probably would not have remained as it was until death. Beetles and other insects have a variety of mechanisms which they can use for righting themselves in these circumstances which, as the writer presumes correctly, must arise often and hazardously.

The most famous mechanism is used by the click beetles (*Elateridae*), which are able to launch themselves into the air by the sudden release of a blunt spine which is kept under pressure in a specialised groove on the venter.

As many readers will have noticed, the click beetle often makes several attempts before it lands on its feet, but its success, given time, is assured.

Other less sophisticated beetle correcting mechanisms include spreading the wings, reaching out with the legs, and rocking the body in a forward-aft or side-to-side motion.

Christopher Starr
Department of Zoology
University of the West Indies
St Augustine, Trinidad and Tobago

Only a minority of beetles possess a body plan that poses such a problem. For example, I have worked with several species of ladybeetle (*Coccinellidae*) in the laboratory, and most are able to right themselves with relative ease.

The species that do find themselves stranded on their backs tend to be the larger varieties that possess strongly convex elytra (the first pair of hardened, protective wings).

Ladybeetles that do become stranded on a smooth surface will eventually unfold their membranous hind wings, which are normally hidden beneath the elytra, and then use these to right themselves. Part of the answer, then, is that very few species become stranded and those that do eventually flip themselves over by means of their hind wings.

Over the long course of evolution it was probably quite rare for beetles developing in temperate forests and grasslands to encounter totally smooth surfaces or bare soil that was devoid of plant litter. Under normal circumstances, grass blades, fallen leaves and plant stems would offer a convenient hold for beetles that happened to become overturned.

The reduced rate of predation and numerous other benefits that are conferred by a hard protective covering, which far outweighs the occasional stranding, has contributed to the enormous evolutionary success of beetles. In terms of both absolute numbers and numbers of species, beetles are the most successful group of animals on the planet.

Tom Lowery
Pest Management Research Centre
Ontario, Canada

I doubt whether the beetle was a healthy specimen that just happened to fall over and was unable to right itself. It is more likely that it was an old, sick or diseased specimen that was nearing the end of its life. When this happens in beetles, they lose a great deal of their mobility and coordination and they become very unstable when walking. They frequently fall over when placed on a hard flat surface and are unable to right themselves.

I have observed this countless times in a number of beetle groups. In fact, while growing up I lived near Milwaukee, Wisconsin, in the USA. We had a fairly large population of *Carabus nemoralis*, which is a ground beetle that was introduced from Europe into the USA. I would frequently find beetles on the sidewalks on their backs. No matter how many times they were righted, they would invariably end up on their backs again, soon to die. I also observed beetles stagger out of the vegetation bordering the sidewalk, only to fall onto their backs. If these beetles were placed on their feet, even in the vegetation, they would stagger about and would fall onto their backs again when they encountered the sidewalk.

So, I suspect that the poor design is really a combination of dying beetles coupled with a smooth, hard surface – one that is not normally found in nature. Considering that roughly one out of every five living creatures is a beetle, and that they occupy virtually every niche and habitat known, I would suggest that beetles are, in fact, very well-designed animals.

Drew Hildebrandt
By email, no address supplied

Does anything eat wasps?

In a recent conversation about food chains, a colleague wondered if anything ate wasps. Someone suggested 'very stupid birds'. Does anyone know any more about this?

Tom Eastwood
London, UK

The lowly wasp certainly has its place in the food chain. Indeed, the question should possibly be 'what doesn't feed, in one way or another, on this lowly and potentially dangerous insect?'

Here are a few that do, the first list being invertebrates: several species of dragonflies (*Odonata*); robber and hoverflies (*Diptera*); wasps (*Hymenoptera*), usually the larger species feeding on smaller species, such as social paper wasps (*Vespula maculata*) eating *V. utahensis*; beetles (*Coleoptera*); and moths (*Lepidoptera*).

The following are vertebrates that feed on wasps: numerous species of birds, skunks, bears, badgers, bats, weasels, wolverines, rats, mice and last, but certainly not least, humans and probably some of our closest ancestors.

I have eaten the larvae of several wasp species fried in butter, and found them quite tasty.

Orvis Tilby
Salem, Oregon, USA

The definitive source on European birds, *Birds of the Western Palearctic*, lists a remarkable 133 species that at least occasionally consume wasps. The list includes some very unexpected species, such as willow warblers, pied flycatchers and Alpine swifts, but two groups of birds are well-known for being avid vespivores. Bee-eaters (*Meropidae*) routinely devour wasps, de-stinging them by wiping the insect vigorously against a twig or wire. And honey buzzards raid hives for food. They are especially partial to bee larvae, but in the UK, wasps, again mostly larvae, also form a major part of their diet.

Simon Woolley
Winchester, Hampshire, UK

I have a photograph taken in my garden, showing a mason wasp having its internal juices removed via the proboscis of a large insect.

Tim Hart
La Gomera, Canary Islands, Spain

In July 1972 I was snorkelling off the Californian island of Catalina. I returned to the east cliff of the island as sunlight was leaving the shore. In a crevice at the base of the cliff I saw a crab holding a wasp, which was still moving.

I took a photograph which shows the right pincer holding part of the wasp while the left pincer carries the wasp's abdomen to the crab's mouth.

The crab did not show any sign that it was startled by the taste of its meal.

Garry Tee
Auckland, New Zealand

Badgers will dig out a wasps' nest and eat the larvae and their food base. During the summer of 2003 I saw an underground nest being demolished by badgers.

Tony Jean
Cheltenham, Gloucestershire, UK

I was once idly observing a wasp crawling round the edge of a water lily leaf in my pond when it paused to drink.

There was a sudden flurry of activity when a frog leapt from its hiding place and swallowed the wasp. The frog did not appear to suffer any ill effects, so I captured another wasp, tossed the hapless creature into the pond and waited. The frog was slow on the uptake, but there was another disturbance in the water and this time a goldfish snapped up the wasp. The fish, too, seemed undisturbed.

My curiosity now thoroughly aroused, I wondered whether the fish could be induced to consume further wasps. For the next hour or so I continued to hunt down luckless wasps and throw them into the pond. Some got away, some were eaten by the fish, and a few were swallowed by the frogs.

John Croft
Nottingham, UK

Returning home late one night I heard the persistent buzzing of a wasp in the kitchen window. It appeared to be struggling around at the bottom of the window, unable to fly properly. A tiny red spider was attached to the underside of its abdomen. The spider must have been some twenty times smaller than the wasp and was positioned where the wasp was unable to mount a counter-attack.

The next morning revealed an empty, transparent wasp exoskeleton.

John Walter Haworth
Exeter, Devon, UK

How do butterflies fly?

Butterflies have a very haphazard flight pattern that no doubt serves them well as a defence against predators. But with their small eyes and small brains how can they see or understand where they are going?

Peter Koch
Le Touvet, Auvergne-Rhône-Alpes, France

Release a nocturnal moth by day and as likely as not a bird will gobble it up. But birds normally take no interest in daytime butterflies, which move about freely, seemingly ignored. Nonetheless, I have seen birds try to catch butterflies on several occasions and each time it was clear why they are normally left alone: they are too quick and manoeuvrable.

It is not true, though, to say butterflies have small eyes. Each compound eye covers most of the side of the head and is huge compared with the size of the insect. They evolved compound eyes because a simple eye with the same light-gathering power would be heavy and cumbersome.

Butterflies also have three small simple eyes, called ocelli, on the top of the head. These serve, so far as is known, at least two purposes. They help the insect maintain a correct flight attitude and they assist in navigation. All day-flying insects have large eyes compared with their nocturnal counterparts, showing the relative importance of vision.

Typically, although not exclusively, a male butterfly will stake out a territory in a sunny spot and display to attract a female, seeing off any other male with the same idea. Initially at least, the female finds a partner by sight, observing the brightly coloured display. Butterflies also find flowers by sight to feed on their nectar.

They do not see the same light spectrum as humans: theirs is shifted into the ultraviolet, while red is invisible, so flowers usually have a UV component.

Humans cannot see the UV reflected from a field of oilseed rape in flower on a bright day, but we can tell it is there from the shimmering, dazzling character of the light.

As for understanding, there is not much room in a butterfly for a brain. The nervous system consists of a series of connected ganglia – collections of nerve cells – with the ganglion in the head merely larger than the others.

Priority has to be given to motor control and reflex behaviour; there is no space for much more. To survive, butterflies have to do four things: feed, mate, find a plant on which to lay eggs and, in certain cases, migrate. Understanding is not necessary.

It should not be forgotten either that butterflies have antennae – quite a lot of 'seeing' is done with these. A male will not be deceived by a predator simulating a female because the true female gives off a pheromone sensed by receptors on the antennae. Flowers, carrion, faeces and rotting fruit can all be found by odour too. This is also how a female finds the correct plant on which to lay eggs. The butterfly confirms the plant's identity by tasting it with her feet when she lands.

Terence Hollingworth
Blagnac, Occitanie, France

How high do butterflies fly?

My four-year-old daughter asked me how high butterflies fly. I was stumped. Can anyone tell us?

Jacque (and Tara) Lawlor
Chelmsford, Essex, UK

Unlike humans, butterflies are not disposed to seeking altitude records. Indeed, they will not fly higher than is strictly necessary in their everyday lives, whether looking for a mate, food or somewhere to lay eggs, avoiding predators or migrating.

Worldwide there are many thousands of species of butterfly, each adapted to its own particular habitat and needs. Some spend their whole lives on a patch of coastal grassland, the larvae feeding on low plants or living in ants' nests, and the adults never flying more than a few feet above the ground. Others spend all their time in the tree canopy many metres above ground level.

Still others are only found on high mountains. So even though they don't actually fly very high above the ground locally, butterflies that live on the mountains of Peru spend their whole lives at altitudes of around 6,000 metres.

Butterflies that migrate tend to fly the highest in general.

The most famous migratory butterfly is probably the monarch (*Danaus plexippus*). These leave Mexico each year and fly north to Canada, albeit taking several generations to get there. Monarchs have been sighted by glider pilots flying as high as 1200 metres. Interestingly, they seem to fly in the same way as a glider, using updrafts to gain sufficient altitude so that they can glide for quite a distance before needing to use energy to climb again.

Europe also has plenty of migratory species. The painted lady (*Vanessa cardui*) makes its way to southern France from north Africa. It has to leave Europe in winter as no development stage of this insect can survive a frost.

To get to France many will cross through the mountain passes of the Pyrenees, which in general lie at about 2500 metres. During late summer and autumn one can observe butterflies drifting southwards. If they encounter a high building, they just fly straight upwards and over it. If they encounter a high mountain range, they will do the same. So you need only to stand for a while on any mountain pass during the migration period to see them coming over either singly or in swarms, flying close to the ground as they travel.

The mountain passes of the Caucasus are higher, while those of the Himalayas are higher still at 7500 metres. I wouldn't be surprised if migratory butterflies could fly straight over Everest if they encountered it in good weather.

However, insects of any kind cannot fly if they are too cold. Butterflies can keep warm to a certain extent by beating their wings, though if they fly too high in the wrong conditions, they may become too chilled to maintain a wingbeat.

On average, the air temperature reaches freezing at an

altitude of just below 8000 metres, suggesting that this would be their physical altitude limit. They might on occasion be carried higher on updrafts, but this surely doesn't count as autonomous flight.

Terence Hollingworth
Blagnac, Occitanie, France

The greatest acknowledged height achieved by migrating butterflies is 5,791 metres, set by a flock of small tortoiseshells (*Aglais urticae*) crossing the Zemu glacier in the eastern Himalayan mountains.

Not only is this an altitude record for butterflies, it is also the highest that any insect has been observed in controlled flight, comfortably exceeding the more frequent altitudes of between 3000 and 4000 metres at which monarch butterflies have been sighted by commercial airline pilots.

Hadrian Jeffs
Norwich, Norfolk, UK

Could bats eradicate malaria?

If there are 250,000 bats in a colony and each one eats 3375 mosquitoes in a night, how long will it take the bats to rid one Texan swamp of malaria? Charles Campbell, a doctor in San Antonio, did the sums and was convinced that bats could consume enough mosquitoes to eradicate the disease from the swamplands around the city. But how could he persuade some of the millions of bats that flew in from Mexico each spring to settle where the mosquitoes were worst? Campbell's answer was the bat tower, a high-rise home to tempt even the most discerning bat. And not only would the tower's tenants transform the malarial mires into wholesome and habitable land, they would provide a steady income. Bat guano was a highly prized fertiliser, and if one bat produced 40 grams of guano a year, then a quarter of a million bats . . .

It seemed so simple. Mosquitoes spread malaria. Bats eat mosquitoes. Set the bats on the mosquitoes and the disease would disappear. Charles Campbell knew about malaria: he was a doctor. He also knew about bats, which he studied in his spare time. There was no doubt that they ate mosquitoes, and lots of them. He had poked through countless pellets of bat guano and totted up the fragments of wings, legs and other indigestible bits to come up with an estimated nightly toll.

At the turn of the twentieth century, malaria was still rife in America's southern states. Each year, there were hundreds of thousands of cases, costing the nation around $250 million. But while parts of Texas were plagued by mosquitoes, the state was also home to huge populations of bats. Late each February, a hundred million Mexican free-tailed bats fly north from Mexico to central Texas, forming immense colonies in caves, old barns and derelict buildings. Campbell's plan was to install bats in purpose-built roosts close to places where mosquitoes bred.

In 1902, Campbell lined some wooden boxes with guano-coated cheesecloth and fixed them in old buildings, under country bridges, and in trees near a cave where millions of bats roosted. After five years and no bats, Campbell conceded that the boxes were a dismal failure. Bats, he deduced, preferred larger homes.

Within months he had built a ten-metre tower at the local experimental farm. It cost him $500. Inside were sloping shelves for the bats to cling to, a large heap of guano to make them feel at home and 'three perfectly good hams with a nice slice cut out of the side of each, exhibiting their splendid quality for the delectation of the intended guests'. But there were no guests. Eventually Campbell resorted to kidnap, capturing 500 bats and incarcerating them in the tower, hoping their squeaks would attract others. By 1910, it was clear that bats were never going to move in. He dismantled the tower and sold the timber for $45.

Disappointed, Campbell left his practice and headed for the wilds of Texas to learn more about bats. He discovered that they don't hunt immediately outside their roost but make a fast getaway to avoid predators that might be lying in wait for them. In April 1911, Campbell built his second tower on the shore of Mitchell's Lake, just south of San Antonio.

'No swamp in the low lands could possibly be worse,' he said. All San Antonio's sewage flowed into the lake, and water seeping from it formed a huge shallow pool – a perfect breeding place for mosquitoes, and the right sort of distance from the tower. In summer great clouds of mosquitoes drove the tenant farmers from their fields around the lake and tormented their animals. Crops went to ruin. Cows stopped producing milk and hens gave up laying. Worse, this was the malaria season. Of the eighty-seven people Campbell examined that spring, seventy-eight had malaria. Each year two, three or sometimes four children died.

Three months later, Campbell returned to check on his tower. Just after dusk, a stream of bats emerged. It took five minutes for all of them to come out. This was promising, but the tower could hold many more bats than that – at least 250,000, perhaps even half a million. Campbell knew of two roosts nearby, one in a derelict ranch house, the other in a duck hunters' shack. If he could evict the bats, maybe they would seek refuge in his tower.

Campbell had tried evicting bats before. Shouting, clapping, even hosing them down with water would dislodge them, but they always returned. He decided to try music. Campbell reasoned that because bats have sensitive hearing, tuned 'to detect the soft sonorous tones made by the vibrations of the wings of mosquitoes', they might be upset by less agreeable noises. 'A brass band suggested itself,' he wrote later. From hundreds of recordings he picked the Mexico City Police Band's rousing rendition of 'Cascade of Roses' 'on account of the large number of reed instruments and some blatant high notes of cornets'.

With the help of a friend with a phonograph, Campbell tested his bat-scarer at the abandoned ranch house. At four a.m. the brassy tones of Mexico City's finest began to belt out of the building. An hour later, the bats began to return from hunting. They circled the house a dozen times, then fled for good. A repeat performance emptied the duck hunters' shack. But had the bats moved into the tower?

They had. On Campbell's next visit the evening stream of bats took two hours to come out.

By 1914, local duck hunters reported a huge reduction in mosquitoes. Campbell visited the farmers and their families again. None had malaria. The farmers told him the clouds of mosquitoes had gone. They could work after dusk and their animals were healthy. Campbell attributed the mosquitoes' disappearance to his bats: 'There was nothing that could have brought about this modified condition except the great increase in the number of bats.'

Impressed, the San Antonio city council made it an offence to kill a bat, and stumped up for a municipal bat roost. In 1917, bats became protected throughout Texas. More towers sprang up – and not just as a way to defeat malaria. From the start, Campbell had promoted the idea of bat towers as a nice little earner. He had weighed a free-tailed bat's daily output and had done the sums. The Mitchell's Lake tower produced about two tonnes a year of high-quality fertiliser that fetched twice the price of the stuff from caves.

If Campbell's bats eradicated malaria, why isn't he one of the state's great heroes? Although malaria did disappear from San Antonio around this time, many doubt the bats had anything to do with it. 'Campbell assumed bats ate mosquitoes but his identification was suspect,' says Thomas Kunz, a bat expert at Boston University. A meticulous study of the free-tailed bat's diet carried out at Gary McCracken's bat lab at the University of Tennessee, Knoxville revealed that on average, 33 per cent of the diet is moths, 30 per cent is beetles and 15 per cent bugs. Dipterans – which include mosquitoes – make up just 2 per cent of the diet.

Stomach analyses and lab tests of guano failed to turn up anything resembling mosquito remains. 'If the bats are eating mosquitoes then it must be a tiny part of the diet,' says McCracken.

Merlin Tuttle, founder of Bat Conservation International in Austin, Texas, isn't so sure. 'The bats' main food is moths but they are highly opportunistic. If there were a large number of mosquitoes then they might eat them.'

Today it's impossible to test whether Campbell's bats helped to eradicate malaria. The land has been drained, most of the mosquitoes have gone and there's no longer malaria in the USA. 'The disappearance of malaria may have been serendipitous,' says Kunz. As Tuttle admits: 'We simply don't know what happened.'

Why do flies like eating dog turds?

To me, this seems horrible. Why don't they get sick like a human would?

Cindy Germond
Sydney, Australia

This is a larger issue than dog turds. Parents worldwide teach children faeces are dirty, messy, bad and yucky. Thus the stuff itself and all the synonyms for it that we learn as children become lodged in the brain as a bad thing. We do not play with excrement. This extends to all excrement from any source.

Through this process we learn from a young age that excrement may harbour 'germs', which is sometimes true, although not all microbes are germs and therefore bad for you. Nonetheless, the lesson we learn is a good one because much gastrointestinal disease is transmitted by the so-called 'faecal-oral' route, making hand-washing after visiting the toilet imperative.

Because I was a paediatric gastroenterologist for almost forty years, I have looked through the microscope at thousands of specimens of human faeces, searching for evidence of maldigestion (undigested fat or muscle), or more commonly inflammation (white and red blood cells). At the microscopic level, excrement is a seething mass of microbes – bacteria, archaea, yeast, and occasionally visible protozoa or fungi all scooting around either actively or by Brownian movement among the debris of digested food particles and cellular cast-offs from the bowel. The microscopic view is almost a work of art as the energetic micro-world recycles debris into life forms that will in turn become food for something further up the food chain.

So to answer the question: flies were never instructed by their parents to avoid faeces, dog or human, and neither were African dung beetles. So they do not object to the odours, hydrogen sulphide, dimethyl sulphide and similar compounds that have been formed by sulphur-metabolising microbes in our colon, or to the appearance of the turd itself. They just see a buffet where much of the digestive work has already been done and they can feed, lay their eggs and relax.

Maybe the *Lord of the Flies* gets to enjoy the biggest dump.

Adrian Jones
Professor Emeritus, Pediatric Gastroenterology and Nutrition
University of Alberta, Edmonton, Canada

Sadly, in our mindless quest for economic growth, humans rarely appreciate the services the natural world provides free of charge and instead we often seek to trash it.

Luckily for us – or perhaps unluckily when it's our favourite item of clothing – insects are less discerning and will eat any kind of organic matter, whether it be fur, dead vegetation or excrement. For their part insects profit from a ready supply of food and energy, while we profit because such waste products, which otherwise would build up and bury us, are recycled relatively rapidly back into the food chain.

As far as faeces are concerned, despite having passed through a digestive tract, they still contain enough organic material to feed the larvae of flies, and often the adult flies as well. Effectively, dog faeces are composed of material which the dog has not digested, just as there is organic matter in our faeces which humans can't digest – think sweetcorn. Flies are able to digest the remainder.

Insects which live in environments with a high microbial load have evolved a resistance to potentially harmful microbes. For instance, they have genes that govern the synthesis of antimicrobial peptides, which, to them, are effectively broad-spectrum antibiotics.

Dogs and humans play host to a great variety of bacteria in the gut which are actually beneficial. The superbug bacterium *Clostridium difficile* may be present in a healthy gut without causing harm but it can cause problems for hospital patients whose immune systems are weak. A controversial way of restoring gut health after an episode of diarrhoea, say, is to inject a sample of faeces from a close relative into the colon.

So it is not necessarily true that people would fall ill eating dog faeces. I would not recommend the practice, however, because one major danger is that they may contain eggs of flat or roundworms of the genus *Toxocara*, which present no threat to insects but which can establish themselves in humans.

Nonetheless, in cities with many dogs, and particularly in winter when insects feed on faeces less, we can unwittingly ingest a significant amount of airborne bacteria from that source.

Terence Hollingworth
Blagnac, Occitanie, France

Where did all the locusts go?

In the summer of 1873, black clouds drifted east from the foothills of the Rocky Mountains towards the newly settled farms of Nebraska, Iowa, Minnesota and the Dakotas. The pioneer families had no warning: the sky went dark at midday, the air filled with a sound like a thousand scissors. Then the clouds fragmented and locusts fell like hail onto crops of corn and wheat. In a few hours, the insects had devoured months of work. Locusts had been invading farms on the American frontier on and off for decades, but the irruption of the mid-1870s entered into legend. Many families gave up farming and fled to the cities. On 26 April 1877, John Pillsbury, the governor of Minnesota, called for a day of prayer to plead for deliverance from the locusts. A few days later, the insects rose up and left as inexplicably as they had come.

When the Rocky Mountain locusts swarmed, they darkened the skies over vast swathes of the western and central USA, from Idaho to Arkansas. The number of insects was mind-boggling: one reliable eyewitness estimated that a swarm of locusts that passed over Plattsmouth, Nebraska in 1875 was almost 3,000 kilometres long and 180 kilometres wide. And they were devastating. 'You couldn't see that there had ever been a cornfield there,' one farmer said after a swarm passed through his land. Yet between these episodes of frantic fecundity, the locusts seemed to disappear.

The Rocky Mountain locust (*Melanoplus spretus*), a big, beefy species of grasshopper, was considered the greatest threat to agriculture in the West. So entomologists tried to learn everything they could about the insects – what triggered them to swarm, what they ate and how they reproduced. But after the spring of 1877, the locusts vanished and never plagued western farmers again. Within thirty years of Minnesota's official day of anti-locust prayer, the insect was extinct. The last live specimen was found by a river on the Canadian prairie in 1902.

No one mourned the loss, and scientific interest in the locust waned. In the 1940s and 1950s, when farmers began to wage war on their enemies with insecticidal chemicals, a few researchers began to speculate about what could possibly have seen off the locust so spectacularly in those pre-pesticide days.

During the disastrous outbreaks of the 1870s, farmers fought back with every tool they could find or invent. They deliberately set their fields on fire. They dragged tar-coated hunks of metal through the ground, hoping to trap locust hatchlings in the sticky goo. Nothing helped much. When desperate pioneer women tried to protect their vegetable gardens by draping blankets over them, the locusts ate the blankets before moving on to the vegetables.

Whatever had done for the locust, it seemed, was some event far beyond the capabilities of nineteenth-century farmers. As the extinction coincided with a time of dramatic environmental change across the West, there were plenty of plausible explanations. Perhaps the locusts had depended on the fires that Native Americans had routinely set to keep the prairies open. Or maybe their most crucial habitat had been shaped by the huge herds of bison that were now all but extinct.

Most standard entomology texts claimed that extreme fluctuations in population, like those that took place when the locusts swarmed, were a sign of a species in trouble, fighting to recover a balance with its environment. The sweeping changes that came with settlement, some scientists suggested, pushed the locusts through cycles of population explosion and collapse, and in the end wiped the species out.

When Jeffrey Lockwood, an insect ecologist at the University of Wyoming, was hired to explore the biology of grasshoppers in 1986, the post-mortem on the Rocky Mountain locust had not gone beyond this sort of general speculation. Lockwood wanted to know more, and he hoped that somewhere there were still a few specimens of the long-vanished locust to study. Among the high peaks of the Rockies in Montana and Wyoming were glaciers where swarming insects had fallen, become immobilised by the cold, and died. Some of these glaciers might still hold frozen remains of the Rocky Mountain locust.

Lockwood and his students spent summers searching in the ice at remote spots high in the Rockies. They began their hunt at Grasshopper Glacier in Montana, hoping it might live up to its name. Sure enough, they found some scattered body parts that might have once belonged to Rocky Mountain locusts, but without whole bodies there was no way to prove these bits had not belonged to some other, still living, species of grasshopper.

'Finally, after four years of fruitless searching, we found the mother lode,' says Lockwood. On Knife Point Glacier in the Wind River Mountains of Wyoming, they recovered 130 intact bodies of Rocky Mountain locusts, the legacy of a swarm that had risen out of the river valleys of western Wyoming in the early 1600s. The antiquity of the frozen insects – confirmed by radiocarbon dating – proved that locusts had irrupted long before European settlers changed the face of the West. The reproductive frenzies, which at times produced enough insects to blanket the entire state of Colorado, were normal events in the history of the locust. Further study of Knife Point Glacier revealed deposits of locust parts throughout the layers of ice, indicating that swarms passed over the mountains at regular intervals during the centuries before the locust's extinction.

To find more clues to what killed off the locust, Lockwood began to scour the scientific literature of the late 1800s. There he found the writings of Charles Riley, an entomologist who had spent much of the 1870s and 1880s searching for ways to kill the locusts.

Riley had mapped what he called 'the permanent breeding zone' of the locust, the territory where mating adults and eggs could be found every summer, regardless of whether the locusts were swarming. For a species that could spread across much of the continent during outbreaks, this home base was surprisingly small. Between swarms, the locusts lived only in the river valleys of Montana and Wyoming, where they buried their eggs in the damp ground along the banks of streams. These fertile spots were the same places the incoming settlers chose to farm.

Riley had experimented with ways to control the number of locusts. Ploughing, he discovered, could push locust eggs so far down into the soil that they would fail to hatch. Flooding the ground where eggs had been deposited also killed many of the young. Riley concluded that agriculture itself – the processes of ploughing and irrigation – were the strongest weapons against the locust. But because less than 10 per cent of the land in the western USA was arable, he doubted that farming would ever have had a significant impact on the locust.

A century after the locust disappeared, Lockwood took Riley's map of locust egg-laying areas in the 'permanent zone' and superimposed it on a map of land under cultivation for corn, wheat or hay in 1880. He found that he had charted the geography of an extinction. In the 1880s, when the locust population had shrunk during an intermission between outbreaks, every corner of its breeding grounds was being farmed. The settlers, Lockwood suggests, had destroyed their nemesis without ever knowing it, simply by ploughing the land and watering their crops. 'The most spectacular "success" in the history of economic entomology – the only complete elimination of an agricultural pest species – was a complete accident,' he says.

How do ants see off pests?

In 1476, the farmers of Berne in Switzerland decided there was only one way to rid their fields of the cutworms attacking their crops. They took the pests to court. The worms were tried, found guilty and excommunicated by the archbishop. In China, farmers took a more practical approach to pest control. Rather than rely on divine intervention, they put their faith in frogs, ducks and ants. Frogs and ducks were encouraged to snap up pests in the paddies and the occasional plague of locusts. But the notion of biological control began with an ant. More specifically, it started with the predatory yellow citrus ant Oecophylla smaragdina, *which has been polishing off pests in the orange groves of southern China for at least 1,700 years.*

For an insect that bites, the yellow citrus ant is remarkably popular. Even by ant standards, *Oecophylla smaragdina* is a fearsome predator. It's big, runs fast and has a powerful nip – painful to humans but lethal to many of the insects that plague the orange groves of Guandong and Guangxi in southern China. And for at least seventeen centuries, Chinese orange growers have harnessed these six-legged killing machines to keep their fruit groves healthy and productive.

Citrus fruits evolved in the Far East and the Chinese discovered the delights of their flesh early on. As the ancestral home of oranges, lemons and pomelos, China also has the greatest diversity of citrus pests. And the trees that produce the sweetest fruits, the mandarins – or kan – attract a host of plant-eating insects, from black ants and sap-sucking mealy bugs to leaf-devouring caterpillars. With so many enemies, fruit growers clearly had to have some way of protecting their orchards.

The West did not discover the Chinese orange growers' secret weapon until the early twentieth century. At the time, Florida was suffering an epidemic of citrus canker and in 1915 Walter Swingle, a plant physiologist working for the US Department of Agriculture, was sent to China in search of varieties of orange that were resistant to the disease. Swingle spent some time studying the citrus orchards around Guangzhou, and there he came across the story of the cultivated ant. These ants, he was told, were 'grown' by the people of a small village nearby who sold them to the orange growers by the nestful.

The earliest report of citrus ants at work among the orange trees appears in a book on tropical and subtropical botany written by Hsi Han in AD 304. 'The people of Chiao-Chih sell in their markets ants in bags of rush matting. The nests are like silk. The bags are all attached to twigs and leaves which, with the ants inside the nests, are for sale. The ants are reddish-yellow in colour, bigger than ordinary ants. In the south if the kan trees do not have this kind of ant, the fruits will all be damaged by many harmful insects, and not a single fruit will be perfect.'

Initially, farmers relied on nests which they collected from the wild or bought in the market – where trade in nests was brisk. 'It is said that in the south orange trees which are free of ants will have wormy fruits. Therefore the people race to buy nests for their orange trees,' wrote Liu Hsun in *Strange Things Noted in the South*, written about AD 890.

The business quickly became more sophisticated. From the tenth century, country people began to trap ants in artificial nests baited with fat. 'Fruit-growing families buy these ants from vendors who make a business of collecting and selling such creatures,' wrote Chuang Chi-Yu in 1130. 'They trap them by filling hogs' or sheep's bladders with fat and placing them with the cavities open next to the ants' nests. They wait until the ants have migrated into the bladders and take them away.' Farmers attached the bladders to their trees, and in time the ants spread to other trees and built new nests.

By the seventeenth century, growers were building

bamboo walkways between their trees to speed the colonisation of their orchards. The ants ran along these narrow bridges from one tree to another and established nests 'by the hundreds of thousands'.

Did it work? The orange growers clearly thought so. One authority, Chhii Ta-Chun, writing in 1700, stressed how important it was to keep the fruit trees free of insect pests, especially caterpillars. 'It is essential to eliminate them so that the trees are not injured. But hand labour is not nearly as efficient as ant power.'

Swingle was just as impressed. Yet despite his reports, many Western biologists were sceptical. In the West, the idea of using one insect to destroy another was new and highly controversial. The first breakthrough had come in 1888, when the infant orange industry in California had been saved from extinction by the Australian vedalia beetle. This beetle was the only thing that had made any inroads into the explosion of cottony cushion scale that was threatening to destroy the state's citrus crops. But, as Swingle now knew, California's 'first' was nothing of the sort. The Chinese had been experts in biocontrol for many centuries.

The long tradition of ants in the Chinese orchards only began to waver in the 1950s and 1960s with the introduction of powerful organic insecticides. Although most fruit growers switched to chemicals, a few hung on to their ants. Those who abandoned ants in favour of chemicals quickly became disillusioned. As costs soared and pests began to develop resistance to the chemicals, growers began to revive the old ant patrols. They had good reason to have faith in their insect workforce.

Research in the early 1960s showed that as long as there

were enough ants in the trees, they did an excellent job of dispatching some pests – mainly the larger insects – and had modest success against others. Trees with yellow ants produced almost 20 per cent more healthy leaves than those without. More recent trials have shown that these trees yield just as big a crop as those protected by expensive chemical sprays.

One apparent drawback of using ants – and one of the main reasons for the early scepticism by Western scientists – was that citrus ants do nothing to control mealybugs, waxy-coated scale insects which can do considerable damage to fruit trees. In fact, the ants protect mealy bugs in exchange for the sweet honeydew they secrete. The orange growers always denied this was a problem but Western scientists thought they knew better.

Research in the 1980s suggests that the growers were right all along. Where mealy bugs proliferate under the ants' protection they are usually heavily parasitised and this limits the harm they can do.

Orange growers who rely on carnivorous ants rather than poisonous chemicals maintain a better balance of species in their orchards. While the ants deal with the bigger insect pests, other predatory species keep down the numbers of smaller pests such as scale insects and aphids. In the long run, ants do a lot less damage than chemicals – and they're certainly more effective than excommunication.

Do mosquitoes get malaria? Do rats catch bubonic plague?

If not, why not?

Year 5, Christopher Hatton School
London, UK

Rats can get quite sick from plague fleas and some will die, but usually not too quickly. Plague-carrying rats are at their most dangerous when they are about to die, because their fleas leave them as soon as they are dead to find new hosts.

The malaria parasite *Plasmodium* does not usually kill its host mosquitoes, though it may take a high enough toll that it is better for the mosquitoes not to get infected.

If we could breed mosquitoes that were resistant to the parasite we might find that they outcompete ordinary mosquitoes, and this might ultimately help get rid of malaria. This kind of strategy would not work with yellow fever, as the mosquitoes that carry the virus responsible for the illness in humans hardly seem to be affected.

Jon Richfield
Somerset West, Western Cape, South Africa

Plasmodium, the parasite that causes malaria in humans, infects mosquitoes. The mosquitoes then transmit it to people when feeding on their blood. As for the plague microbe, *Yersinia*, it blocks the gut of the flea that transmits it. As a result, when an infected flea feeds on the blood of a human or rat, it will regurgitate some blood containing the microbe and so spread the germ to a new host.

To address the question directly, the important thing to note is that being infected with a microbe or other parasite does not necessarily cause disease, because it is often in the interest of the microbe to cause no harm to its hosts.

However, the mosquitoes that transmit *Plasmodium* are affected by it, as the parasite grows in their salivary glands. Such infection can reduce the ability of the salivary glands to function and thus the viability of the mosquitoes.

A related parasite called *Theileria*, transmitted between cattle by ticks, can damage the gut and salivary glands of the ticks, and can even kill them in the laboratory. Epidemiologists make a point of studying the extent of such effects under natural conditions.

Alan R. Walker
Edinburgh, UK

Where are all the Argentinian ants?

On a recent summer trip to Ushuaia in Patagonia, Argentina, we saw no ants. This troubled us so much that we ended up actively searching them out, with no success. Is there a southern – and indeed a northern – limit to the range of ants or were we just looking in the wrong places?

Andrew and Bronwyn Lumsden
Murrays Run, New South Wales, Australia

In searching around Ushuaia, the questioners found one of the few places on land where ants do not occur naturally, although there is the possibility that adventive species – that is, non-native and non-established ants – survive in houses. It's just too cold and wet there. Other places where you can look in vain are Antarctica, the sub-Antarctic islands (although an adventive species was found once in an abandoned whaler's hut on Kerguelen), the Falkland Islands, the high Arctic, Iceland and the upper slopes of high mountains.

Edward O. Wilson
Department of Entomology Harvard University
Cambridge, Massachusetts, USA

Are insects in the Caribbean smarter?

During summer, if I drive my car in Europe the windscreen is splattered by dead insects. However, I can drive for months in the Caribbean without any insects striking the screen. Are Caribbean insects smarter?

Erik Blommestein
Trinidad and Tobago

It isn't a matter of Caribbean bugs being smarter but of what the local conditions are like or even something as simple as the road size. If you drove along a minor country road in Europe or along an urban highway in the Caribbean, for example, you would probably experience windscreen splatter of the kind you describe.

Many insects are drawn during the evening to any source of light, even if it is only faint, and so may fly towards car windscreens. Insects would not be similarly enticed if you were driving at noon. So the number of insects hitting your screen could be affected by different light levels in Europe and the Caribbean. To make a sensible comparison we need more details, including which species perished.

Jin Xiao
National University of Defense Technology
Changsha, China

The effect could be a consequence of differences in road infrastructure in Europe and the Caribbean. European roads are mostly wide and asphalted, and allow high-speed driving, whereas Caribbean roads are not always so welcoming to the fast driver.

In Europe, traffic and asphalted roads generate and retain heat, which is usually appreciated by insects. In the Caribbean, the difference in temperature on- and off-road is not so great, meaning the road environment may be less attractive to insects. Also, you would probably have to drive more slowly in the Caribbean, giving the insects a chance to escape or [be] swept over the car by the airflow.

Gael Canal
Chaville, Île-de-France, France

You haven't experienced the full fury of bug splatter unless you have driven in Florida in early spring or late summer. The culprit is *Plecia nearctica*, the love bug. The name is well chosen because the insects fly together in pairs. The male locks the end of his abdomen with that of the female, and when she starts to fly the male arches upside down over her and flies with her. In a swarm you seldom see a love bug without a partner flying upside down above it, though this arrangement makes for a very low flying speed.

Love bugs can swarm above roads in such numbers that they almost form a cloud. When you hit an infested section of highway, it can completely obscure your windscreen in one quick blast, as well as the radiator and headlamps. If the black remains are not removed from the paintwork, they can corrode the finish in a day or two.

At the height of the love bug season, many drivers fix a screen to the front of their car to make cleaning them off easier. In the north of Florida, washing stations have been installed along the main highways so drivers can clean the bugs off their vehicles.

William Joseph
UK

Why are flies so noisy?

Why do flies that enter your house in summer do so in such an ostentatious way, with loud and frenetic buzzing? Surely it would be better for their survival if they didn't? Is there a purpose to this display?

Greta Bowman
Brighton, East Sussex, UK

The noisiest flies are blowflies, especially the bluebottle *Calliphora vicina*, which beats its wings around 150 times per second. This generates air vibrations equivalent to 150 hertz; musically speaking it is a D, below middle C. They are noisier than the housefly, *Musca domestica*, because their bigger wings disturb more air. Flies sound louder indoors because there is less background noise to drown them out and their buzzing may reflect off solid surfaces. Their buzzes sound more mechanical than musical because they lack the sweeter harmonics.

Flies clearly evolved with peskiness in mind – forgive my anthropomorphism. They enter without knocking. They aim at your head for their fly-by. They vomit on windows, and are impossible to swat. Then to be really cussed, they taunt you by loitering out of reach and parading about your ceiling. But we might not be asking this question were our hearing beyond their beat frequencies.

The buzz is part and parcel of the very wing beats that initiate flight, so it would be impossible for flies to dispense with buzzing without having to adopt different aerodynamics.

Len Winokur
Leeds, UK

Flies forage constantly, mostly by smell, so they search an area thoroughly before moving on. They treat a house as they would any enclosed area, such as a cave, and when they have finished they look for a way out by using light intensity. In homes this often brings them into contact with glass, which confuses them because they have not evolved to understand it. What amazes me is that a fly can fly full-tilt into a pane of glass without apparently injuring itself.

Since their predators, mostly reptiles and birds (and sometimes pet dogs), tend to hunt by sight and not sound, flies don't care how much noise they make.

Tony Holkham
Boncath, Pembrokeshire, UK

Some flies, such as the irritating blackflies of the family *Simuliidae*, which buzz around your head outdoors on still days, thankfully do not enter buildings. Houseflies such as *Musca domestica*, blowflies such as *Calliphora vomitaria*, and greenbottles of various *Lucilia* species, have no such scruples.

Most people do not realise, I imagine, that much of insect behaviour is regulated by odours. A mosquito finds its victim primarily by detecting body odours and chemicals in breath. The flies described are similarly attracted, not to people, but to chemical signals from food. The blowflies and greenbottles coming into my kitchen make straight for the food remnants in my cat's dish.

If I forget to put it in the fridge during warm weather – the dish, not the cat – it is soon speckled with white patches of eggs. Since the primary food source of such flies is carrion it is hardly surprising that they also do not hesitate to enter dark spaces where a sick animal may have taken shelter and died.

During the colder months the same flies will seek out a sheltered corner indoors where they can hibernate, perhaps behind that piece of peeling wallpaper in the top corner of the room. If a cold house or room warms up, the heat often spurs hibernating flies into activity, although they often then fly around in a drunken, sluggish manner.

I have on occasion shooed flies out of the open house door into the cold only to see them realise their mistake and make a smart U-turn back into the warm.

A house has no significance to a fly, other than as a source of heat, shelter and food. Humans have been around a few hundred thousand years at most while insects have been on Earth at least 400 million years. A few flies trapped or swatted inside houses are irrelevant to their survival as a species.

Terence Hollingworth
Blagnac, Occitanie, France

How are ants able to survive so much?

I have been amazed to see ants emerge seemingly unharmed after being zapped in the microwave, usually after hitching a ride on my coffee cup. They seem to run around quite happily while the microwave is in operation. How can they survive this onslaught?

Judith Kelly
Darwin, Northern Territory, Australia

The answer is quite simple. In a conventional microwave, the waves are spaced a certain distance apart, because that is all that is needed to cook the food properly. The ants are so minute that they can dodge the rays and so survive the ordeal.

Li Yan
Norwich, Norfolk, UK

The phenomenon that the ants take advantage of is that microwaves form standing waves within the oven cavity. So in some places in the oven space, the energy density is very high, whereas in others it is very low. This is why most ovens have turntables to ensure that cooking food is heated evenly throughout.

This standing wave pattern can be observed by putting a static tray of marshmallows in the microwave, and heating for a while. The result will be a pattern of cooked and uncooked marshmallows. The standing wave pattern, however, varies according to the properties and position of any material within the oven, such as a cup of water.

The ant will experience this pattern as hot or cold regions

within the oven and can thus locate the low-energy volumes. If the ant is in a high-field region, its high surface-area-to-volume ratio allows it to cool down more quickly than a larger object while it searches for a cold spot.

It is a common myth that microwaves are too big to heat small objects. The fallacy of this has been demonstrated by chemists such as myself who employ microwave heating in their work. Certain types of catalyst consist of microwave absorbing particles – typically of submicron size – dispersed throughout an inert support material. There is convincing evidence that microwaves are capable of heating only the tiny catalyst particles.

A. G. Whitaker
Heriot, Borders, UK

There is very little microwave energy near the metallic floor or walls of the oven. The electromagnetic fields of microwaves are 'shorted' by the conducting metal, just as the amplitudes of waves in a skipping rope, swung by a child at one end but tied to a post at the other, are reduced to nothing at the post.

An ant crawling on the rope could ride out the motion near the post, but might be thrown off nearer the middle. For a quick demonstration of this, place two pats of butter in a microwave in two polystyrene coffee cup bottoms, one resting on the floor, the other raised on an inverted glass tumbler. Be sure to place a cup of water in the microwave as well. On heating, the raised butter will melt long before the butter on the floor.

Charles Sawyer
Camptonville, California, USA

How do gnats fly in the rain?

How is it possible for gnats to fly in heavy rain without being knocked out of the air by raindrops?

L. Pell
Uffington, Oxfordshire, UK

A falling drop of rain creates a tiny pressure wave ahead (below the raindrop). This wave pushes the gnat sideways and the drop misses it. Fly swatters are made from mesh or have holes on their surface to reduce this pressure wave, otherwise flies would escape most swats.

Alan Lee
Aylesbury, Buckinghamshire, UK

The world of the gnat is not like our own. Because of the difference in scale, we can regard a collision between a raindrop and a gnat as similar to that between a car moving at the same speed as the raindrop (speed does not scale) and a person having only one thousandth the usual density – for example, that of a thin rubber balloon of the same size and shape. A balloon is easily bounced out of the way, and would burst only if it was crushed up against a wall.

Tom Nash
Sherborne, Dorset, UK

How do spiders spin such huge webs?

In a chalet in the Alps which had no obvious draughts, I noticed spider threads spanning horizontally from wall to wall. How did they do it?

Marcello Rebora
London, UK

Air currents carry a silk line from a spider 's spinneret until it reaches a solid object. Because of its sticky nature, the line attaches to whatever it encounters. Draughts or convection currents sufficient to carry the line would exist even in a room that was hermetically sealed. For example, windows allow sunlight to penetrate, heating a patch of floor. Air in contact with the floor would warm and expand. Being less dense, it would rise, to be replaced by cooler, denser air, initiating the required air movement.

The spider scurries along the first line – called a bridge line – spinning a stronger second thread. It continues making return journeys until the line is sufficiently strong. The rest of the web follows.

Mike Follows
Willenhall, West Midlands, UK

Can insects get fat?

Walt Malker
New York City, USA

If, by getting fat, the questioner means obese, the answer is no. All insects undergo some sort of metamorphosis, passing through larval stages before becoming an adult. The adult, or 'imago', stage is relatively short-lived and very often adult insects do not feed at all. Mayflies (of the order *Ephemeroptera*) and many silk moths (of the family *Saturniidae*) are some examples. They neither have the time nor the inclination to feed and get fat.

Those imagos that can feed are constrained by their inflexible exoskeleton. They have no means to expand this to take on excess fat. Incidentally, carcasses of death's head hawk moths that have been stung to death are often found in beehives, where they have made a vain attempt to feed. Their proboscis is too short to enable them to extract nectar from flowers, but long enough for them to consume honey if the bees in the hive would let them.

The larval stages are similarly constrained. To grow they have to moult their exoskeleton periodically. Internal fluid pressure splits the exoskeleton and the insect expands into a new one it has grown underneath, which remains flexible just long enough to accommodate their increased size. They cannot keep doing this indefinitely; each species is limited to a predetermined number of moults. If they find abundant nourishing food, they will go through their moults quickly. If the opposite is the case, or if their food contains limited nourishment, they will take a relatively long time to finish growing. Either way, once the moults are completed, they begin metamorphosis into the adult stage. They do not continue to get fatter and fatter and, in fact, the reverse can be true – those larvae unable to find sufficient food may begin metamorphosis early, skipping one or two moults. This produces a normal, if somewhat smaller than average, imago.

However, insects do store up a great deal of fat at the larval stage. Silk moths are generally large insects. Because they do not feed as adults they must have considerable fat reserves to enable the males to track down a female and mate and, in the case of the females, to produce and lay a large number of eggs while sustaining their metabolism for a week or more. But this is the normal situation; they are not in any way obese.

Terence Hollingworth
Blagnac, Occitanie, France

Insect life cycles do not lend themselves to the concept of obesity as we know it. Most species accumulate food as fast as they can, but once they have enough, they enter their next stage of life, or reproduce and die. There are exceptions, but few can afford to get too fat – anything that interferes with their bodily function prevents reproduction – so what they cannot use, they dump. For example, sap-sucking insects get far too much sugar from plant sap, but instead of becoming uselessly fat, most dump the excess as honeydew or convert it into waxy armour. Using hormones to prevent insects from maturing may make them larger and fatter, but prevents their breeding.

Still, healthy insects do accumulate some fat. Their internal 'fat bodies' are special organs crucial for storage, hormonal control, metabolism, growth, overwintering, fuel for travelling, yolk production for eggs and so on. Accordingly, many insects, such as locusts and termites, though not technically obese, are prized as fat-rich foods. As you may have seen on TV, termite queens of most species accumulate huge fat stores to support their role as virtually continuous egg-laying machines.

Jon Richfield
Somerset West, Western Cape, South Africa

With my colleague James Marden I have described (among other symptoms) infection-associated obesity in a dragonfly species. Infected dragonflies show an inability to metabolise fatty acids in their flight muscles and so build up lipids in their thorax, leading to a 26 per cent increase in thoracic fat content. The suite of symptoms caused by this infection includes decreased flight performance and reproductive success in male dragonflies.

Ruud Schilder
School of Biological Sciences University of Nebraska
Lincoln, Nebraska, USA

If moths are nocturnal, why do they fly towards light?

Asked by BBC Radio 5 Live listeners, UK

Moths are not exclusively nocturnal. Many species are active by day as well, and many are exclusively diurnal.

Insect behaviour is almost entirely instinctive, as the famous ninteenth- and twentieth-century entomologist Jean-Henri Fabre showed us. He noted that a digger wasp of the genus *Sphex* would sting a caterpillar and carry it, paralysed, to the mouth of the burrow she had just excavated and deposit it at the entrance. She would then enter the burrow, presumably to check that no unwanted occupant had taken up residence there in her absence, emerge again, collect the caterpillar and take it inside to lay her egg and close the burrow. After observing this, Fabre began to move the caterpillar a little distance from the hole once the wasp was inside. The wasp would collect the caterpillar and then repeat the inspection process.

Fabre was never able to break this routine. The wasp had evolved to behave this way and it was impossible for her to reason out anything different for this unusual set of circumstances. In the same vein, moths have undergone millions of years of evolution without ever having to deal with artificial lights at night. The few thousand years or so during which we have presented them with the problem is too short for them to have evolved appropriate behaviour.

Richard Dawkins, in his book *The God Delusion* (Bantam Press, 2006), presents us with the problem of moths drawn to a candle flame. His solution is the old glib explanation that the moths are trying to use the flame to navigate, mistaking it for the moon. The idea is that a moth sets its course according to the position of the light, so it will have to keep turning towards it to maintain the same relative heading, and the path it will take will be a spiral leading inevitably into the flame. This explanation does not tell us, however, why it is that in many species only males are thus attracted, and in a few, only females.

What is more, if moths need to navigate, they must be from a migrating species. Yet most of the time such moths are not migrating. Indeed, most species do not migrate at all and thus have no need of navigation. Moreover, all groups of insects display the same behaviour: flies, wasps, hornets, mayflies and caddises are all drawn inexorably towards flames, although many of these insects are normally diurnal and the majority rarely, or never, migrate.

So are they navigating to find food or a mate? At night in summer, male moths use scent to decide their heading, not light. Total cloud cover makes no difference to their behaviour. They move into and across the wind, hunting for telltale pheromones which will lead them to a female or for the scent of flowers to enable them to feed. Females, for their part, stay still until after mating and then go looking for the scent of plants that their larvae can feed on in order to lay their eggs on them. They don't need the moon, stars or candles to do this.

I have spent thousands of hours sitting by light traps observing insect behaviour and I feel, for the most part, it is pure accident that they stumble upon the light. Many can be seen to fly straight past without deviating one iota. Others fly into the lighted area, land and stay still, as they would if it were daytime. Different species seem to have different sensitivities, the most sensitive ones alighting the furthest from the light. Still other moths circle the light, never bumping into it.

Those that do fly towards the light often do so in a wild and confused manner. They seem disorientated and confused by a bright light rather than attracted to it. Just like Fabre's interference with the caterpillar, the light seems to trigger inappropriate behaviour, because the insects have no mechanism to deal with it.

There is only one type of observed behaviour that seems appropriate, and that is moths moving towards a lighted window at night. Any insects trapped in a room will fly to the window. Instinct tells them that in order to escape from a confined space they have to fly to where it is lightest, and even at night it is lighter outside than inside a cave or a cavity where the moth has been hidden during the day. The moth finds its way out by flying towards that light. So a moth would not be able to distinguish between a bright light source and an open space.

It could well be this mechanism that is operating when moths fly to bright lights; however, the many other factors listed above seem to counter this view.

Terence Hollingworth
Blagnac, Occitanie, France

I have seen this question asked many times and answered in as many ways, but I have yet to find an answer as attractive as that which I first read in Ian McEwan's novel *Atonement* (Jonathan Cape, 2001), later a major movie, which is set in the 1930s and 1940s. Incidentally, despite its more historical setting in the novel, I believe the theory originates from

1972 and Henry Hsiao, a biomedical engineer. Put simply, nocturnal moths fly towards dark places and with only simple light-sensing apparatus they perceive the area behind and beyond any point light source, such as a light bulb, as being the darkest around. Sadly, this idea does not seem to be shared by entomologists and has never been confirmed experimentally.

Rob Jordan
Cross, Somerset, UK

Do spiders drink water?

Sarah Cassidy
Banchory, Aberdeenshire, UK

Yes, spiders do drink water. In the wild, most will drink from any available source such as droplets on vegetation or the ground, and from early morning or evening dew that has condensed on their webs. For those kept in captivity, it is a good idea to provide a fresh water source such as a small bottle cap or damp sponge for smaller species, or a small dish for larger species such as tarantulas.

Incidentally, spiders' need to quench their thirst seems to have given rise to the myth that they live in drains. When a spider is in a building, an excellent source of water is droplets left from taps and showers around the plugholes and sink edges. Needless to say, spiders remain trapped in the sink or bath because the sides are too slippery or steep for them to climb, hence their tendency to head for the drain as a means of escape.

Sean Lenahan
Cartmel College, Lancaster University, UK

Like all other animals, spiders require a regular intake of water. Different species use different methods to quench their thirst. For example, the whistling spider, found in the desert, covers its 1-metre-long burrow with a thin layer of silk to keep it humid. Dew or the occasional raindrop is captured using a low, silk-covered mound near the entrance. Many other species, such as the wolf spider, opt for a much simpler strategy by drinking dewdrops in the morning. Some spiders even ingest nectar.

Paul Peng
Nakara, Northern Territory, Australia

Many spiders, such as the common garden spider, will devour their web first thing in the morning. In doing this, they consume the water that has condensed as dew droplets on the web. Other spiders, such as the whip spider, can use their pincers to take water into their mouths.

The black widow or the red back do not drink water at all. They get all the fluid they need from the juice sucked out of their prey. Tarantulas, on the other hand, like to drink water droplets that have collected on nearby leaves and foliage.

There are some creatures, including mammals, that do not drink. The name koala is derived from the Aboriginal word 'no drink'. Koalas get the fluids they need from eating the leaves of plants such as the smooth-barked eucalyptus.

Louise Lench
London, UK

A few winters ago, I watched a spider just outside my kitchen window as a small snowflake landed on its web. Normally a spider does not react to the presence of something in its web unless it struggles, so I was surprised to see the spider run to the snowflake. By the time it arrived, the snowflake had melted into a droplet of water and the spider gave every appearance of drinking it. The spider's head was against the droplet, and the droplet dwindled away to nothing.

Norman Paterson
Anstruther, Fife, UK

The Australian naturalist Densey Cline once reported a remarkable case of a spider drinking water.

She awoke to find the shrivelled body of a dead huntsman spider lying on her bedside table, and in her glass of water was an astonishingly long parasitic worm. She speculated that the mature parasite required water to complete its life cycle and had driven the infected spider to the nearest source of water by inducing a terrible thirst.

Kate Chmiel
Clifton Hill, Victoria, Australia

When an insect is changing inside its cocoon, and has turned to slush, is it alive?

Madeleine Cooke (aged 7)
Ryde, Isle of Wight, UK

In many metamorphosing insects, the majority of the cells in the body of the pupa do break down and turn to mush, but there are clusters of cells that remain intact. These cells feed on the mush, divide, and go on to develop the legs, eyes, wings, antennae and so on that we see in adult insects.

It is almost as though the mush is the yolk and the cluster of cells is the embryo of a new egg. In some rare cases, such as fungus gnats, this new embryo can split to form multiple 'twin' adults from a single larva. This is called polyembryony.

Martin Harris
Australia

An insect undergoing metamorphosis is alive regardless of what state its body may be in. For one thing, the individual cells are alive and are growing and dividing in a coordinated manner to form the organs of the new adult insect. An insect, or any other organism, could not be dead at one stage of its development and alive at a following stage, because the death of an organism is always irreversible.

However, the death of a multicellular organism such as an insect must be defined separately at different levels of organisation: the intact body; the organs and tissues; and finally the individual cells. The body cannot survive without organs and cells, but the latter two groups can survive without a body. If you squash a cocoon the larva inside will be killed, but many of its cells will remain alive, at least for a while. Therefore, a multicellular organism can be killed by destroying its highest level of organisation, while leaving most of its organs and cells alive. If that were not the case, there would not be such possibilities as human organ transplantations or cell cultures.

Aydin Orstan
Germantown, Maryland, USA

What is meant by alive? Is the ball of cells that make up the human blastocyst alive? It cannot breathe, think or feel pain. But it is alive, like the mush of insect cells in a cocoon. The cells that make up both structures are metabolising, dividing and responding to their environment – all hallmarks of life.

Roger Morton
Dunlop, Australian Capital Territory

Can bushes kill?

While taking a break on a country bike ride in Australia, I saw an unfortunate insect impaled on the thorn of a low bush. We'd had strong winds in the days beforehand and I can only assume that the insect was blown onto the thorn, which had penetrated the open wing casing before impaling the body. What are the chances of an event like this occurring?

Paul Worden
Portland, Victoria, Australia

There are examples of impalement of insect species, most often scarab beetles, on spines or other sharp parts of plants and on barbed wire in the zoological literature in Australia, but only rarely.

It is very unlikely that the wind alone could have impaled the elytron – the modified, hardened forewing of beetles which covers their softer flying wings. That would be like an insect collector attempting to pin a beetle by throwing it at the pin. The elytron is part of the hard exoskeleton and would almost always deflect the glancing blow of a pin or spine. The chances of impalement through an elytron while in flight would appear to be very remote because the flying beetle holds its elytra open, at a wide angle to the body, and they would hinge back towards the body if touched on the outer side by a spine.

A more likely scenario is that the strong winds blew down a twig or branch to which the beetle was clinging and that the extra momentum of beetle plus branch impaled it. Similarly, strong winds might have caused one branch to thrash against another on which the beetle was clinging.

Ian Faithfull
Extension Support Officer,
Catchment and Agriculture Services
Carrum Downs, Victoria, Australia

Dung beetles often impale themselves on the barbs of wire fences while in full flight. In New South Wales I quite frequently see barbs on which insects are impaled.

Toshi Knell

Nowra, New South Wales, Australia

Actually, this event is quite likely in certain areas. The poor beetle probably did not get there by chance, but rather because it was put there.

In Victoria, the culprit is likely to have been a grey butcher-bird (*Cracticus torquatus*). These predatory birds, which are about the size of a small dove, eat large insects, small mammals and other birds. They skewer their prey on thorns to hold it while they eat. Sometimes they will impale the prey and leave it for a snack later. Suitable bushes near nesting sites can be festooned with victims, including poultry chicks.

There is a record of a butcher-bird returning to its nest to find its three chicks dead after a spell of cold rain. The bird took them from the nest and hung them in its nearby larder, returning to eat them a few days later.

Shrikes, which are common throughout Eurasia, Africa and North America (and also sometimes known as 'butcher-birds', although they are not related) have similar habits.

Rob Robinson

British Trust for Ornithology

Thetford, Norfolk, UK

Why don't insects fly in straight lines like most birds do?

Their flight patterns seem chaotic and often circular in motion. Surely it is inefficient for them to fly in such a random way?

Mike McCullough
Preston, Lancashire, UK

Birds seldom fly in straight lines, in fact, and any self-respecting ornithologist can identify many species by flight pattern alone. That said, birds tend to fly fairly directly to a target in sight, and more complex journeys tend to follow efficient paths based on learned topography.

Insects' flight is much more varied. Their flight patterns partly compensate for their typically poor eyesight, by allowing them to gather more visuospatial information.

When insects do have a path in mind, many fly directly enough, as anyone who has been stung can attest. In threat display, a bee buzzes wildly, but a true attack is like a projectile from a pea-shooter. In searching, scent-following, mating assembly, territorial patrol and so on, insect flight patterns are not straight, but they do resemble corresponding vertebrate behaviour.

Differences between, say, insects' apparently chaotic swarms and skeins of large birds do not reflect their psychology but rather their size difference, which affects wing motions and the benefits of flying in fixed formation.

Flocks of small birds such as starlings or queleas can look very locust-like.

Jon Richfield
Somerset West, Western Cape, South Africa

Judging insect flight depends on the scale you choose. If you watch a metre of flight and you see the insect travelling in a straight line you might conclude it does this all the time. But if you observe a kilometre of flying the picture is quite different.

When insects are searching, their flight is an intermittent combination of long and short-distance steps where each step differs from the previous one by a small angle. Small steps are the most frequent, while very long steps are rare. This is an optimal searching strategy known as 'Lévy flight'. It seems chaotic but it is not; the angles and the distance follow well-defined statistical distributions. In the case of bees, they fly in amazingly straight lines for kilometres once they know where a food source is. When they aren't flying 'randomly', a straight line is the best way to go.

Octavio Miramontes
Mexico City, Mexico

Insects do not have vision as sharp as that of mammals or birds. The insect compound eye is more attuned to movement and so it cannot precisely position distant objects. As a result, insects tend to take a rather wobbly flight path to navigate to a particular object.

Additionally, many insects navigate using scent. Take the case of a parasitic wasp that is seeking a caterpillar in which to deposit its eggs. In order to locate the caterpillar, the wasp needs to balance the odour signals received by its two antennae. This necessitates a rather wobbly flight path to 'lock on' to the source of the scent. When an insect is close to the object it is seeking, this wobble starts to reduce and eventually the insect becomes capable of very precise, rapid actions, such as those of a dragonfly catching a prey insect when both are in mid-flight.

The 'random' movement is a simple product of insect sensory systems. Humans rely on vision to navigate the world – just try locating an open bottle of perfume in a room with your eyes closed.

Peter Scott
School of Life Sciences, University of Sussex
Brighton, UK

I suspect evolution wasted no time in eliminating those insects which fell easy prey to bats, birds, frogs and the like because of their use of efficient and direct – but predictable – flight patterns. Presumably those insects with variable patterns are those that survived until today.

Peter Tredgett
Llanbedrog, Gwynedd, UK

Creature Curiosities

Spiders on speed

Spiders on marijuana are so laid back that they weave just so much of their webs and then . . . well, it just doesn't seem to matter anymore. On the soporific drug chloral hydrate, they drop off before they even get started.

A spider's skill at spinning its web is so obviously affected by the ups and downs of different drugs that in 1995 scientists at NASA's Marshall Space Flight Center in Alabama thought spiders could replace other animals in testing the toxicity of chemicals. Different drugs had varying effects on the average arachnid addict. On benzedrine, a well-known upper, house spiders spun their webs with great gusto, but apparently without much planning, leaving large holes. On caffeine they seemed unable to do more than string a few threads together at random.

The more toxic the chemical, the more deformed the web. NASA researchers had hoped that with help from a computer program they would be able to quantify this effect to produce an accurate test for toxicity.

Wasps — sniffer dogs with wings?

In 2006, after three ten-second training sessions, Glen Rains' crack team of sniffers was ready for anything. They could be co-opted into the hunt for a corpse. They might join the search for a stash of Semtex or a consignment of drugs. Or they could have the more tedious job of checking luggage at the airport. Whatever the assignment, their role was the same: to pick up a scent no human nose could detect and pinpoint its source. These recruits to the fight against crime were smaller, cheaper and more versatile than a sniffer dog, and more sensitive than an electronic 'nose'. They were wasps.

Insects have exquisitely sensitive olfactory systems. Their

antennae are covered with microscopic sensors that can detect the faintest odour. Some are also remarkably quick learners. So it was hardly surprising that they aroused the interest of the military and security services, police and customs, all badly in need of ultra-sensitive, flexible and portable odour detectors. Insects obviously have the right stuff, but could they use it to sniff out smells they would never encounter in nature – a hint of explosives, say, or a whiff of cocaine? And if so, would it be possible to make a practical device that harnessed their skills?

Enter Wasp Hound, a hand-held odour detector with a team of little black wasps as its sensor. Developed by Rains, a biological engineer at the University of Georgia, his colleague Sam Utley and Joe Lewis, an entomologist at the US Department of Agriculture's Agricultural Research Service in Tifton, Georgia, the device was only a prototype, but the team had high hopes for it.

Did this mean the days of the sniffer dog were numbered? Probably not. 'We don't see insects as a replacement for dogs,' said Rains. 'But they do have lots of advantages. They cost pennies to raise. They don't need special handling, and because they are so quick to train you can have them on call, ready to learn a new smell whenever you need it.'

IS IT A . . .?

BIRDLIFE

Do magpies prefer semi-skimmed milk?

I get two bottles of milk delivered to my house each day. One, containing whole milk, has a silver foil top, whereas the other, containing semi-skimmed milk, has a silver top overprinted with red stripes. Based on observation over several years, the local magpie population will often try to peck at and remove the striped top but hardly ever attack the plain silver foil top. Have other readers observed magpies or other birds being so discerning, and is there a scientific explanation for it?

Barry Chambers
University of Sheffield, UK

There are two potential explanations. Firstly, the magpies in your garden may be showing some form of aversion to the silver bottle-tops. Birds often show unlearned aversions to food of certain colours, but these tend to be colours that are associated with toxic insects, such as the black and yellow stripes of wasps or the red and black spots of ladybirds. While this could be the case with your magpies, I doubt it because the red-striped tops would appear to be more reminiscent of the colour patterns of toxic insects than the silver tops.

The theory I favour is that your birds know what's good

for them! Insectivores such as magpies need a diet rich in protein with lower levels of both carbohydrates and fat. Just like in humans, high-fat diets can cause magpies to suffer from high levels of cholesterol and all the medical problems associated with that. Birds are excellent judges of the toxin and nutrient content of the food they eat and by choosing to drink the semi-skimmed milk over the whole milk they get the benefit of a high-protein food source without the costs associated with eating fatty foods.

Maybe magpies could teach us all a thing or two about healthy eating.

John Skelhorn
The Institute of Neuroscience
Newcastle University, UK

In my youth, we used to have ordinary milk, with a red top, and creamy Guernsey milk, with a silver top, delivered to our home. The blue tits always attacked the creamier Guernsey bottles. After a couple of years, the dairy changed the cap colours to silver and gold, respectively. The birds learned the new colour code in about two weeks.

Your magpies are either stupid or fashionable – assuming, of course, there is a difference.

Alan Calverd
Bishop's Stortford, Hertfordshire, UK

Why is bird poo white?

Local birds tend to eat little black insects. So how come they void themselves on me from a great height with a white and annoyingly conspicuous product?

M. Rogers
Great Hockham, Norfolk, UK

It is a common misconception that the white droppings produced by birds are faeces. In fact, they are urine. Birds excrete uric acid rather than urea because it is an insoluble solid. This way they avoid wasting water when urinating – just one of their adaptations for a good power-to-weight ratio.

Guy Cox
University of Sydney, Australia

The white material that comprises the droppings of birds, and indeed many reptiles, is their urine.

The more primitive vertebrates excrete toxic nitrogenous waste relatively directly, having masses of water at their disposal with which they can dilute substances such as ammonia.

However, birds and reptiles – at least lizards and snakes, with whose droppings I'm very familiar – are different. It would appear that the conversion of their toxic nitrogenous waste products into a relatively insoluble one that can then be formed into a paste was an evolutionary adaptation. This enabled them to lead a terrestrial rather than aquatic life, and even to live in ecological niches where water is scarce.

In such niches it is particularly important not to have to

find extra water with which to dilute toxic waste products and flush them from the system, so birds and lizards solved this by evolving to produce a paste of insoluble and relatively non-toxic uric acid.

Interestingly, birds that consume a lot of roughage with their diets, such as the heather-eating grouse and ptarmigan, produce droppings that are very similar to guinea-pig faeces. Only here and there among the droppings is it possible to make out the telltale white patches of their urine, so copious is their production of faeces.

Philip Goddard
By email, no address supplied

Your previous correspondents omit one fact, oviparity. The evolution of insoluble excreta has nothing to do with a 'good power-to-weight ratio' or the ability to 'live in ecological niches where water is scarce'.

It evolved because all birds and many reptiles begin their life inside an egg. Even heavy egg-laying amniotes that live in water as adults, such as penguins and crocodiles, must survive this early phase without poisoning their shelled enclosure with any water-soluble metabolites.

Örnóflur Thorlacius
Reykjavik, Iceland

They do so from a great height because from a lower height it's just too easy to hit the target – no challenge at all. The deposit needs to be white so that, from said great height, they can see where it lands and who it hits.

S. B. Taylor
Canterbury, Kent, UK

What makes gulls dance?

While sitting on a bench beside a local green, I noticed a gull performing an excellent version of Riverdance. Then it stopped and scrutinised the grass around its feet. This sequence was repeated for about fifteen minutes. I assume the gull was trying to attract worms to the surface with its rhythmic dance. Was it? If so, how does the strategy work?

Danny Hunter
Dublin, Ireland

Yes, like many species of birds, some gulls have learned the earthworm-raising trick. Earthworms stay underground during the day unless flooded out by rainwater or alarmed by ground vibrations that suggest the approach of a mole. Just jab a garden fork into earth well populated with earthworms and some will pop out to avoid the little creature in black velvet.

Different birds have different techniques. Blacksmith plovers, rather than hunting earthworms, flush out grasshoppers, caterpillars and moths by tickling short grass with a trembling foot held forward.

Gulls that have learned the trick stamp for earthworms. Similarly, I have seen thrushes stamp by hopping hard with stiff legs. Once I was startled to see a red-winged starling watching an olive thrush's technique attentively, then having a go itself. It did get a worm or two, but its technique was faulty, with long, loping leaps instead of jerky thumps, so it did not scare enough worms and soon gave up. Or maybe it just didn't like the flavour of those earthworms it had caught.

Antony David
London, UK

The gull was indeed trying to get worms to surface. Underground, the rhythm of the gull's feet sounds much like rain.

Earthworms like to surface during rain because it enables them to move around overground without drying out – this is impossible when it is dry. By tricking the earthworms, the gulls get an easy meal. The gulls may have learned this trick from watching other gulls, or may have inherited the behaviour.

Laura Still
Devon, UK

I was sitting on Henley Beach in South Australia recently, watching a gull 'puddling' the sand at the water's edge before inspecting the water for any food items it might have disturbed. It seemed to be doing quite well for itself.

It appears that the gull seen by the questioner was applying successful food-gathering behaviour that evolved in one environment to another. This does make sense when one considers that gulls originate not in marine environments, as is frequently supposed, but in moorland ones. Presumably the behaviour evolved in environments that contained bogs, where the puddling behaviour would work successfully in damp ground some way from the water's edge.

The vital question, though, is whether the gull was successful in drawing up worms, or anything else edible, to the surface.

Graham Houghton
Aldgate, South Australia

How does a mother duck recognise her chicks?

I was watching a duck and her eight chicks walking in a line across the grass. All of a sudden a couple of other chicks entered the group. The mother duck immediately weeded out the stranger chicks and sent them on their way. To us they looked identical, so just how did the mother duck achieve her feat? Is it just that animals are exquisitely sensitive to visual differences between members of their own species? Or was the mother duck relying on non-visual information as well and, if so, what?

Byung O Ho
San Jose, California, USA

Chasing away non-descendant young is called 'kin discrimination' and is often considered less efficient in birds than in other animals. However, eider ducks have been reported to discriminate against ducks that are not part of their family unit. Coots have also been seen to do the same thing, but neither species seems to use appearance as the way to recognise their young.

Many birds use acoustic recognition and can identify each other's voices. Swallows, finches, budgies, gulls, flamingos, terns, penguins and other birds that live in larger flocks do this. Odours can also play a role in determining how some birds recognise each other.

In ducks, sound seems to be the principal method of recognition: they have been fooled into returning to the wrong nest, only to be greeted by a portable cassette player rather than their ducklings.

The ability to recognise their own young saves colony-living birds from expending energy in raising someone else's offspring. It also stops ducklings running the risk of aggression from adults if they beg food from the wrong ones. Natural selection favours individuals who know who they are talking to.

Waterfowl have long been thought to be unable to keep track of their own young. They have been seen to lose their own ducklings to another parent, or to mistakenly accept and care for non-descendant ducklings. This has been put down to the fact that birds do not generally have a central family unit.

Ducks do behave in a different way towards their own ducklings, though. Parents sometimes favour their own offspring over non-descendant young, as with the duck in this question, or they may tolerate or encourage the ducklings to mix. Consequently, some provide what is called alloparental care, a form of adoption. This is seen when a duck is able to increase the chances of survival of her own offspring by accepting non-descendant ducklings into her entourage. Her own ducklings might be better off because the risk of any individual being eaten by a predator is lower if it is part of a bigger group. To improve the advantage even more, the non-descendant ducklings may be positioned at the edge of the brood, further away from parents. This has been seen in Canada geese; the adopted goslings were noted to generally potter further away from their adoptive parents than the biological offspring, and therefore not as many survived.

Jo Burgess
Department of Biological Sciences
Rhodes University, Grahamstown
South Africa

Why do some birds stand on one leg?

Alexander Middleton
Moorooka, Queensland, Australia

It has been proposed that the reason that flamingos stand on one leg is so ducks don't swim into them as often! The most likely answer, though, has to do with energy conservation. In cold weather, birds can lose a lot of heat through their legs because the blood vessels there are close to the surface. To reduce this, many species have a counter-current system of intertwined blood vessels so that blood from the body warms the cooler blood returning from the feet. Keeping one leg tucked inside their feathers and close to the warm body is another strategy to reduce heat loss.

I imagine the converse is true in hot climates – blood in the legs will heat up quickly, so keeping one leg close to the body will reduce this effect and help the birds to maintain a stable body temperature.

Another factor in long-legged birds is that it may require significant work to pump blood back up the leg through narrow capillaries. Keeping the leg at a level closer to the heart may reduce this workload.

It is also worth remembering that birds' legs are articulated differently to ours; what looks like the knee is in fact more like our ankle. Many birds have a mechanism to 'lock' the leg straight, so for them it is much easier to stand for hours on end on just one leg – on numerous occasions I have seen birds take off, and even land, on one leg.

Rob Robinson
Senior Population Biologist
British Trust for Ornithology
Thetford, Norfolk, UK

Do birds have a slow metabolism?

During migration the ruby-throated hummingbird (Archilochus colbris) *tanks up with a few drops of nectar for the last time on the northern shores of the Gulf of Mexico. It then flies non-stop for at least 800 kilometres to reach the shores of the southern Gulf. Can anyone calculate the metabolic fuel efficiency of these birds that fly so far on so little, and how does this compare to a human?*

Martin Bradfield
Lohhof, Bavaria, Germany

This question may involve some erroneous assumptions. Before departing for the Yucatán peninsula, a hummingbird spends weeks gorging on arthropods and does not merely consume 'a few drops of nectar'. It puts on enough fat to nearly double its weight: a female can grow from 3.2 grams to around 6 grams, and can barely get airborne. When, after anything up to twenty two hours, it reaches its destination, it will weigh around 2.7 grams, having consumed the fat and often some muscle tissue. Many do not complete the trip.

The average metabolic rate for the black-chinned hummingbird is 29.1 ± 6.3 kilojoules per day. A man, metabolising energy at the same rate, would have to consume twice his weight in meat a day, or 45 kilograms of glucose, and his body temperature would rise to over 400°C.

Lanny Chambers
St Louis, Missouri, USA

The basal metabolic rate is the measurement of how much oxygen an organism uses when at rest. Just the fact that the hummingbird has a very high ratio of body mass to surface area gives it a basal metabolic rate that is twelve times as high as a pigeon's and 100 times that of an elephant.

The metabolism of the ruby-throated hummingbird is much lower when in torpor than in flight. To go from torpor to an active state takes it about an hour. The heart rate rises from 50 beats per minute to 500, and its temperature from 10°C to 40°C. When in full flight its heart pumps at 1260 beats per minute and its wings beat 50 to 200 times per second. And this is still more energy-efficient than a human taking a walk.

Ruby-throated hummingbirds belong to a group of birds known as passerines, which have three toes facing forwards and one toe back. Passerines tend to have a metabolic rate that is as much as 70 per cent higher than either non-passerine birds or mammals. Their muscles are made up of about 35 per cent mitochondria with densely packed cristae – infoldings of their inner membrane – which makes them at least twice as efficient as human mitochondria.

To achieve the same efficiency, humans would have to have muscles composed of 70 per cent mitochondria. And even then the muscles could not work because there would be too few myofibrils in them.

While wintering in Mexico, the bird doubles its weight by building up considerable fat reserves to use on its long journey north. In fact, most ruby-throated hummingbirds travel along the edge of the Gulf of Mexico, eating a little along the way, but some do take the short cut from Florida to Yucatán using their fat reserves and catching gnats along the way.

A number of key factors make the ruby-throated hummingbird so efficient. Its pectoral muscles are red meat, which is rich in the oxygen-carrying protein myoglobin; the muscle has a high capillary-to-fibre ratio, giving it a good blood supply; it has an energy-rich diet of nectar that is stored as fat; it can eat nectar from any species of flower so it does not waste time looking around for a particular source, and will also eat any insects; its tongue is fringed so that the nectar is effortlessly drawn up by capillary action; and it operates at high temperatures to make its metabolic reactions more efficient.

It has been calculated that this species eats as much as three times its own weight in a day, during which it is awake for sixteen hours on average. This is the equivalent of an 83-kilogram human eating 125 kilograms of hamburgers every day, or 1,335,000 kilojoules.

Mike Ball
Gorinchem, South Holland, Netherlands

Why do birds sing as dawn breaks?

Indeed, why do some sing at dusk? And for what reasons do they eventually stop?

Eva Sanz
Tarragona, Catalonia, Spain

The main function of most birdsong is long-distance communication, either to mark territory or to be sociable. As such it is largely intraspecific; blackbirds sing to impress blackbirds, not buntings. In contrast, social vocalisation, such as coordinating group activity, largely occurs at short range during active flight or foraging, or when settling down for the night or preparing to take flight as a flock.

Like any form of communication, birdsong bears an energy cost and requires channel capacity, which is limited largely by background noise and the quality of the medium, in this case air. In the mornings and evenings the air tends to be still, which reduces competing noise. It is also cooler at low altitudes, which favours transmission of clear sounds. Also, few birds forage at dusk, so in terms of energy use that is an economical time to perform.

Because much birdsong is territorial, it is practical for each species to sing at fixed times to avoid wasting energy on talking when no one is listening or when other species are competing for air time. Ideally, that male blackbird would be saying: 'If you are a male, keep off, or else! But if you are a female, let's get together.' Later, when other species are singing, he can go off to catch the early, deaf, worm.

Antony David
London, UK

Birds sing more at dawn and dusk than at other times because that is when they can hear more birds singing. Frequently at these times the wind drops and a temperature inversion forms – this is a layer of warmer air above cooler air. This changes the way in which sound is carried through the air, refracting sound waves back towards the ground that otherwise would have dissipated in the air. The upshot is that sound is carried further at dawn and dusk.

Thus, if a bird devotes most of the energy it spends on singing to those times, it is heard by the widest possible audience. Of course, there are birds that can be heard singing at any time of day, but even these will tend to sing more at dawn and dusk when there is a temperature inversion.

Nigel Depledge
Spennymoor, Durham, UK

How do birds' eggs all hatch on the same day?

Patrick Casement
London, UK

In the majority of bird species synchronous hatching is the norm. This is achieved simply: the parent does not commence incubation until the clutch of eggs, however large, is complete.

Most birds will lay an egg each day until an appropriate trigger indicates that the clutch is complete. This is why I can collect an egg a day from each of my hens – they keep laying until incubation is triggered by some stimulus. This could be the feel of a full clutch against the brood patch on a bird's belly, but there is also some endogenous control factor. For example, one of my hens may turn 'broody' – she will cease laying and commence incubation on only one egg if that is all I have left her with.

In some other groups of birds – notably owls, raptors and cormorants – incubation commences when the first egg is laid, leading to sequential hatching, with the first chick gaining a significant advantage over later ones. This is very noticeable in barn owls, where five or six eggs laid at daily intervals lead to the oldest chicks being almost a week older than the youngest. This strategy ensures maximum chick survival in species with an unpredictable food source. In years of plenty all the offspring get enough food, but in years when food is less readily available the oldest, largest chicks survive and dominate the younger ones – which almost inevitably perish and may be eaten by their siblings.

This strategy sees its most elegant expression in the Cain-and-Abel syndrome, which is manifested particularly well in eagles. The first egg hatches two to three days before the second and, when food is scarce, there seems to be a degree of inevitability in the way the older chick persecutes its younger sibling to the point of death.

Norman McCanch
Canterbury, Kent, UK

There are billions of birds worldwide, so why is it that you rarely, if ever, see a dead one?

Maurice Boland
Radio Marbella, Spain

The fate of a bird carcass, or indeed that of any other animal, depends largely on the cause of death. If the bird was killed by a predator, it will probably be eaten immediately and there will be nothing left to see except for the odd feather.

Animals that are sick generally hide somewhere quiet and isolated. So the bodies of birds that die from disease or from old age will be in hard-to-get-to places and will most probably be eaten by ants and other insect scavengers before you chance across them.

The only case remaining, and the main reason we humans see dead animals, is when the animal is killed accidentally by something with no interest in eating it. The most obvious example of this is road kill, although most birds are fairly fast-moving and agile, and so are less likely to be hit by cars than slower-moving ground animals. However, on a recent trip to central Queensland my parents saw several dead wedge-tailed eagles by the side of the road. These majestic birds, which can have a wingspan of 2.5 metres, are unfortunately rather cumbersome and slow to take off. Therefore, when they land on roads, they are often killed by vehicles.

Ironically the eagles had been attracted to the roadside by the carrion of other road kill.

Simon Iveson
Cleveland, Queensland, Australia

For each dead bird there are, fortunately, plenty of burying beetles of the family *Silphidae* that will promptly fly to the scene from a considerable distance away, attracted by the enticing smells. So tough is competition for these tasty morsels that sometimes beetles carry mites on their bodies that upon arrival promptly alight on the corpse and start ridding it of the eggs of blowflies or other faster-eating scavengers. This buys the beetles some time while they scrape away under the bird, which will soon sink into the ground and disappear.

Once they have located a corpse, a couple of beetles will soon mate and start preparing the nest for their offspring. With the help of mouth and anal secretions they make a 'brood mass' with the bird's flesh and tend it so that by the time their eggs hatch it will still be in pristine condition for their larvae.

During the early stages of the larvae's life the parents will feed them with regurgitated bird flesh much as birds do for their nestlings. This form of parental care is very rare among such non-social insects, but burying beetles will look after their larvae until they are ready to pupate in the soil. By that stage there will be not much left of the bird at all.

Maria Fremlin
Essex, UK

Dead birds accumulate in some areas, such as the edges of lakes, estuaries and shorelines – this is why avian flu surveyors focus on places like this – and also along roads, where millions of birds die.

Remains are mostly quickly scavenged, but in areas where game birds are released in large numbers, roads are often covered in dozens of casualties. Sometimes these can be useful for stocking one's larder, but one must question the ecological impact (and perhaps ethics) of large-scale autumn releases of more than twenty million pheasants and red-legged partridges in the UK.

I recommend a visit to the estate roads of Scotland for an instructive 'dead bird' experience.

Ian Francis
Alford, Aberdeenshire, UK

Do pigeons sweat? If not, why not?

Class 3L, Hungerford Primary School
London, UK

Only mammals have sweat glands – so no, pigeons do not sweat. Nor do mammals such as cats, whales and rodents, which have lost most or all of their eccrine sweat glands – the ones that we use in shedding heat – while birds never developed them. In sweat-free mammals the kidneys deal with the excretory functions of sweat, and flushing or panting is how they cool themselves down. As another example of evaporative cooling, an over-hot cat not only pants but also moistens its fur with saliva.

Bird skins are dry. However, as birds, including pigeons, have body temperatures that are generally several degrees higher than those of mammals, they do not need the same capacity to lose heat.

When they do need to lose heat, they can raise their down feathers to cool the skin by ventilation. To conserve heat, they flatten them. Beyond this, panting through open beaks causes evaporative cooling: hence the Afrikaans expression: 'So hot the crows are yawning' ('So warm dat die kraaie gaap').

Finally, on very hot days, many kinds of birds, including pigeons, enjoy a bath.

Byron Wilson
Dublin, Ireland

Storks, cormorants and vultures indulge in urohydrosis: they literally wet and thus cool themselves by urinating down their legs. Because birds do not urinate and defecate separately, everything comes out together, which makes bird droppings very watery. The heat required to evaporate this liquid from the surfaces of the legs cools the blood, carried close to the surface of the legs by a network of veins.

Before condemning these birds for their unappealing party trick, it is worth adding that bird droppings contain uric acid, making it an effective antiseptic – very useful for vultures that spend a lot of their time trampling over rotting carcasses.

Mike Follows
Willenhall, West Midlands, UK

Why do seabirds sound so different from land-based birds?

In New Zealand one of our radio stations broadcasts native birdsong each morning. It is obvious that seabirds have a much harsher screeching sound than the more melodious bush and land-based birds. In fact, I can usually tell a bird's habitat simply by the sound it makes. Why is there such a difference, and is it the same throughout the world?

John Finlayson
Maungaturoto, Northland, New Zealand

Birdsong indeed varies by habitat type because the habitat has a profound effect on how these long-distance signals are transmitted. To minimise habitat-induced degradation, the acoustic adaptation hypothesis predicts that birds living in dense forests will have slower and more tonal calls, while those living in more open habitats will have faster-paced and buzzier calls.

The effect is most pronounced when comparing contrasting habitat types, such as very open and very closed ones. Other factors, including the songs of species competing for acoustic space and the songs produced by closely related species, can also play a role.

Daniel T. Blumstein
Department of Ecology and Evolutionary Biology
University of California
Los Angeles, USA

The subject is more complex than the question suggests. The South African bush hosts croaking corvids, harmonising antiphonal shrikes, shrieking parrots, raucous francolin, swizzling weavers and tweeting wagtails.

Calls seem to be adapted to distance, noise, obstacles, habit and competition. The most elaborate singers inhabit open bush, where their song can convey complex information over long distances. In thick bush, only deep ventriloqual notes such as those of the ground hornbill carry for any distance. White-eyes foraging among dense leaves cheep softly, keeping flocks together at short range.

Even the apparently unsophisticated croaks, screams and yarps of seabirds vary in complexity and carrying power according to their habits and individual circumstances. When calling through wave noise over long distances they tend to screech shrilly, whereas when they are intimate they are quieter.

Details vary, but the fundamental principles of auditory information encoding and transfer seem inescapable.

Jon Richfield
Somerset West, Western Cape South Africa

Why do birds never fall off their perches when sleeping?

Do they, in fact, sleep?

Graeme Forbes
Kilmarnock, Ayrshire, UK

Birds have a nifty tendon arrangement in their legs. The flexor tendon from the muscle in the thigh reaches down over the knee, continues down the leg, round the ankle and then under the toes. This arrangement means that, at rest, the bird's body weight causes the bird to bend its knee and pull the tendon tight, so closing the claws.

Apparently this mechanism is so effective that dead birds have been found grasping their perches long after they have died.

Anne Bruce
Girvan, Ayrshire, UK

Yes, birds do sleep. Not only that, but some do it standing on one leg. And even more surprising, may be hypnotised into sleep at will. My myna bird was.

If you wish to try it, bring your eyes close to the cage, and use the hypnotist's 'your eyes are getting heavier' principles (not spoken) on your own eyes. The bird will gradually lift one leg up under its belly, tucking its head under its wing and falling into a deep sleep.

What's more, most pet bird owners know that all you need to do to make your pet fall asleep is to cover the cage with a blanket to simulate night.

David Leckie
Haddington, East Lothian, UK

Birds do sleep, usually in a series of short 'power naps'. Swifts are famous for sleeping on the wing. Since most birds rely on vision, bedtime is usually at night, apart from nocturnal species, of course.

The sleeping habits of waders, however, are ruled by the tides rather than the sun.

Some other species are easily fooled by artificial light. Brightly lit city areas can give songbirds insomnia. A floodlit racetrack near my home gives an all-night dawn effect on the horizon, causing robins and blackbirds to sing continuously from 2 a.m. onwards. Unfortunately, I don't know whether it tires them out as much as it does me . . .

Andrew Scales
Dublin, Ireland

Why are most eggs egg-shaped?

Max Wirth
Bowness-on-Windermere, Cumbria, UK

Eggs are egg-shaped for several reasons. First, it enables them to fit more snugly together in the nest, with smaller air spaces between them. This reduces heat loss and allows best use of the nest space. Second, if the egg rolls, it will roll in a circular path around the pointed end. This means that on a flat (or flattish surface) there should be no danger of the egg rolling off, or out of the nest. Third, an egg shape is more comfortable for the bird while it is laying (assuming that the rounded end emerges first), rather than a sphere or a cylinder.

Finally, the most important reason is that hens' eggs are the ideal shape for fitting into egg cups and the egg holders on the fridge door. No other shape would do.

Alison Woodhouse
Bromley, Kent, UK

Most eggs are egg-shaped (ovoid) because an egg with corners or edges would be structurally weaker, besides being distinctly uncomfortable to lay. The strongest shape would be a sphere, but spherical eggs will roll away and this would be unfortunate, especially for birds that nest on cliffs. Most eggs will roll in a curved path, coming to rest with the sharper end pointing uphill. There is in fact a noticeable tendency for the eggs of cliff-nesting birds to deviate more from the spherical, and thus roll in a tighter arc.

John Ewan
Wargrave, Berkshire, UK

Eggs are egg-shaped as a consequence of the egg-laying process in birds. The egg is passed along the oviduct by peristalsis – the muscles of the oviduct, which are arranged as a series of rings, alternately relax in front of the egg and contract behind it.

At the start of its passage down the oviduct, the egg is soft-shelled and spherical. The forces of contraction on the rear part of the egg, with the rings of muscle becoming progressively smaller, deform that end from a hemisphere into a cone shape, whereas the relaxing muscles maintain the near hemispherical shape of the front part. As the shell calcifies, the shape becomes fixed, in contrast to the soft-shelled eggs of reptiles which can resume their spherical shape after emerging.

Advantages in terms of packing in the nest and in the limitation of rolling might play a role in selecting individuals which lay more extremely ovoid eggs (assuming the tendency is inherited) but the shape is an inevitable consequence of the egg-laying process rather than evolutionary selection pressure.

A. MacDiarmid-Gordon
Sale, Cheshire, UK

Do penguins' feet freeze?

Why do Antarctic penguins' feet not freeze in winter when they are in constant contact with the ice and snow? Years ago, I heard on the radio that scientists had discovered that penguins had collateral circulation in their feet that prevented them from freezing but I have seen no further information or explanation of this. Despite asking scientists studying penguins about this, none could give an answer.

Susan Pate
Enoggera, Queensland, Australia

Penguins, like other birds that live in a cold climate, have adaptations to avoid losing too much heat and to preserve a central body temperature of about 40°C. The feet pose particular problems since they cannot be covered with insulation in the form of feathers or blubber, yet have a big surface area (similar considerations apply to cold-climate mammals such as polar bears).

Two mechanisms are at work. First, the penguin can control the rate of blood flow to the feet by varying the diameter of arterial vessels supplying the blood. In cold conditions the flow is reduced, when it is warm the flow increases. Humans can do this too, which is why our hands and feet become white when we are cold and pink when warm. Control is very sophisticated and involves the hypothalamus and various nervous and hormonal systems.

However, penguins also have 'counter-current heat exchangers' at the top of the legs. Arteries supplying warm blood to the feet break up into many small vessels that are closely allied to similar numbers of venous vessels bringing cold blood back from the feet. Heat flows from the warm blood to the cold blood, so little of it is carried down the feet.

In the winter, penguin feet are held a degree or two above freezing – to minimise heat loss, whilst avoiding frostbite. Ducks and geese have similar arrangements in their feet, but if they are held indoors for weeks in warm conditions, and then released onto snow and ice, their feet may freeze to the ground, because their physiology has adapted to the warmth and this causes the blood flow to feet to be virtually cut off and their foot temperature falls below freezing.

John Davenport
University Marine Biological Station
Millport, Ayrshire, UK

I cannot comment on the presence or absence of collateral circulation, but part of the answer to the penguin's cold feet problem has an intriguing biochemical explanation.

The binding of oxygen to haemoglobin is normally a strongly exothermic reaction: an amount of heat (ΔH) is released when a haeomoglobin molecule attaches itself to oxygen. Usually the same amount of heat is absorbed in the reverse reaction, when the oxygen is released by the haemoglobin.

However, as oxygenation and deoxygenation occur in different parts of the organism, changes in the molecular environment (acidity, for example) can result in overall heat loss or gain in this process.

The actual value of DH varies from species to species. In Antarctic penguins things are arranged so that in the cold peripheral tissues, including the feet, DH is much smaller than in humans.

This has two beneficial effects. Firstly, less heat is absorbed by the birds' haemoglobin when it is deoxygenated, so the feet have less chance of freezing. The second advantage is a consequence of the laws of thermodynamics. In any reversible reaction, including the absorption and release of oxygen by haemoglobin, a low temperature encourages the reaction in the exothermic direction, and discourages it in the opposite direction. So at low temperatures, oxygen is absorbed more strongly by most species' haemoglobin, and released less easily. Having a relatively modest DH means that in cold tissue the oxygen affinity of haemoglobin does not become so high that the oxygen cannot dissociate from it.

This variation in DH between species has other intriguing consequences. In some Antarctic fish, heat is actually released when oxygen is removed. This is taken to an extreme in the tuna, which releases so much heat when oxygen separates from haemoglobin that it can keep its body temperature up to 17°C above that of its environment. Not so cold-blooded after all!

The reverse happens in animals that need to reduce heat due to an overactive metabolism. The migratory water-hen has a much larger DH of haemoglobin oxygenation than the humble pigeon. Thus the water-hen can fly for longer distances without overheating.

Finally, foetuses need to lose heat somehow, and their only connection with the outside world is the mother's blood supply. A decreased DH of oxygenation by the foetal haemoglobin when compared to maternal haemoglobin results in more heat being absorbed when oxygen leaves the mother's blood than is released when oxygen binds to foetal haemoglobin. Thus heat is transferred into the maternal blood supply and is carried away from the foetus.

Chris Cooper and Mike Wilson
University of Essex, Colchester, UK

Creature Curiosities

Hen night

A 1958 report from Germany threw an unexpected sidelight on the behaviour of hens. The agricultural research institute at Würzburg had been studying what, if any, were the harmful effects of wines produced from hybrid grapes.

Germany, like France and Switzerland, was plagued

with a sizeable output of wine of extremely low quality and dubious potability. In the experiments, hens, spiders and goldfish were all used, and the hens in particular consumed astounding quantities of wine. They were divided into three groups, the controls being given water only and the others pure wine and hybrid wine respectively. Each hen was given half a pint of liquid a day.

Incredibly, sixteen hens got through 600 pints of red wine in four months, seemingly in radiant health. These sixteen did not, however, include any of the hens that drank hybrid wine – these died early in the study. The fate of the spiders and goldfish was not disclosed.

Landing strip

Here is a possibly apocryphal story sent to us by a friend who got it from a friend who . . . etc.

It concerns a student at the Massachusetts Institute of Technology who went to the Harvard football ground every day for an entire summer wearing a black and white striped shirt. He would walk up and down the pitch for ten to fifteen minutes throwing birdseed all around him, blow a whistle and then walk off the field. At the end of the summer, the Harvard football team played its first home match to a packed crowd. When the referee walked on in his black and white strip and blew his whistle, hundreds of birds descended on the field and the game had to be delayed for half an hour while they were removed.

The student, so the story goes, wrote his thesis on this, and graduated.

No pushover

It's official: penguins don't fall over backwards when aircraft fly overhead.

British pilots came back from the 1982 Falklands War with stories of penguins toppling over. Concerned that increasing air traffic might endanger wildlife, a team led by Richard Stone of the British Antarctic Survey spent five weeks in 2001 watching a thousand king penguins on South Georgia. After numerous overflights by two Royal Navy Lynx helicopters, 'Not one king penguin fell over,' Stone told Reuters.

THE LIFE AQUATIC

FISH AND OTHER UNDERWATER ANIMALS

Do fish get thirsty?

Jack Bennett
By email, no address supplied

Well yes, at least some of them do, so long as we leave aside the subjective human feeling of 'thirst'. There is also a substantial difference between fish in seawater and freshwater, and we need to consider the possibility of the thirsty shark.

Bony fish, known as teleosts, have a salt concentration in their bodies that is not dramatically different from that of land-dwelling vertebrates. This means that the teleosts of the sea – marine fish – live in an environment with a much higher salt concentration than is present in their blood. Their relatives in freshwater are in the opposite position.

Water tends to move along concentration gradients through water-permeable biological membranes like those that shield most organisms from their environment – a process known as osmosis. Therefore, marine fish, which have a low salt concentration compared with that of seawater, will constantly leak water through their body wall – especially through the thin and permeable gill epithelia. To replenish this lost water, marine fish need to drink, so it would be easy to argue that they become thirsty. The surplus salt they ingest by drinking seawater is excreted by specialised cells located in the gills.

Freshwater fish, on the other hand, are unlikely to become thirsty. Because they live in a more dilute environment, they have the opposite problem: water flows inwards and dilutes their blood. The freshwater fish therefore need to excrete excess water, which they do in much the same way we do, via a dilute urine. So we can see that marine fish get thirsty and drink, while freshwater fish avoid drinking but pee a lot.

Finally, sharks, dogfish, rays and skates – which are cartilaginous rather than bony and are called elasmobranchs – are also marine fish (with a few Central and South American freshwater exceptions). Although the concentration of inorganic salts in their blood is not dramatically different from that of marine teleosts, they have little or no osmotic gradient between blood and seawater. This is because they retain organic molecules instead, the main ones being urea (carbamide) and trimethylamine oxide (TMAO). In this way, the cunning sharks avoid an osmotic water flow from their body surfaces, and may not be very thirsty.

Stefan Nilsson
Professor of Zoophysiology
University of Gothenburg, Sweden

. . . and what about dolphins?

So what about water-dwelling mammals such as dolphins and whales. Do they get thirsty? And if they do, how do they drink?

Daniel Gough
Glasgow, UK

Dolphins and whales do not drink. Just as we humans cannot use saltwater as our source of water, neither can marine mammals. This is because they would need to ingest more freshwater than the seawater they consume in order to excrete the salt it contains.

Much of their water comes from fish and squid, which can contain more than 80 per cent water by mass. They can also obtain water through metabolising fat. In order to reduce their water loss they have similar internal designs to those of desert-dwelling mammals, including a long loop of Henle in the kidney nephron.

As well as internal adaptations, marine mammals did away with sweat glands to stop any water loss through sweating. Instead, they use their surroundings to cool down.

Matthew Tranter
Newcastle-under-Lyme, Staffordshire, UK

Marine mammals certainly are less prone to thirst than land-dwelling mammals; for one thing, they have no need to sweat. They do not swallow any more saltwater than they can help, though. Unlike seabirds and turtles, they lack special

Could you mummify a toad?

Toads have always been linked to the supernatural. Stone-like themselves, they were thought capable of surviving within solid rock. Even today, every encyclopedia of the 'unexplained' offers sober accounts of entombed toads. Yet despite all the stories, there's just one tangible example. It's a mummified toad nestled within a hollow flint – and the most famous specimen in a provincial English museum. Its sterling credentials are tarnished by just one thing: the man who presented it to science was Charles Dawson, the discoverer of Piltdown man.

By Victorian times, entombed toads had become a simmering controversy that refused to go away, rather like psychic phenomena or UFOs today. Predictably, the media leapt on each new example: the living toad discovered two metres down in bedrock beneath a cellar in the Lincolnshire town of Stamford or the one found in a block of limestone during the excavation of a new waterworks for Hartlepool, in England's far north. Even *Scientific American* ran a story about a silver miner called Moses Gaines who came across a tiny but plump toad lodged in a boulder.

It was all too much for the scientific establishment. When a live frog, allegedly released from a lump of coal mined in Monmouthshire, was shown at the 1862 International Exhibition in London, the letters pages of *The Times* were filled with furious demands for its removal.

So, on 18 April 1901, when the Linnean Society met ir

London, its august fellows must have been intrigued to see the exhibit brought by one Charles Dawson. A solicitor from Sussex and a fellow of the Geological Society, Dawson was a charming man with wide-ranging interests in geology, natural history and antiquities. In his bag he carried a curious find – 'a hollow flint nodule which had been picked up on the downs at Lewes, and which on fracture was found to contain the desiccated body of a Toad'.

A small hole visible at one end must have provided the way in for a very young toad, the fellows agreed. Helpfully, Dawson had unblocked the hole, removing the chalk that must have subsequently filled it. Once inside the rock, he suggested, the toad lazily remained, content to dine on such insects that found their way in, until, too big to escape, it died entombed.

So it was curiosity, laced with sloth, that ensured the toad's fate – a salutary lesson from nature. Even better, Dawson's specimen offered a rational explanation for all those troublesome tales of toads living for millennia buried in stone. Now, at last, the truth was clear: the flint was ancient – the hollowed remains of a Cretaceous sponge – but the toad was a recent arrival. Thanks to Dawson, the mystery was solved.

A little over a week later, the toad made its second public appearance before the Brighton and Sussex Natural History Society. If anyone doubted the stories about toads being found alive in rocks, here was the explanation, Dawson proclaimed. 'Toads when small will often creep into holes in rocks and hollows in trees, and in these situations they may find sufficient food; being slothful in their habits, and capable of existing upon but little food or of abstaining from it for a long time, they are apt to remain in their snug quarters and content themselves with what insects &tc may come to them.' Besides, he reckoned, flies might have been quite plentiful, perhaps lured inside by the 'fetid and acrid exudations' from the toad's skin, or the sound of the soft scratching of its claws.

To safeguard the specimen's future, Dawson gave it to his friend Henry Willett, a wealthy Brighton collector, who included it in his generous gifts to the town's new Booth Museum of Natural History. In a small way, Dawson had made his mark in the annals of science.

And so the matter might have rested, had not Dawson gone on to 'find' the Piltdown skull. In 1953, this fossil was exposed as a deliberate fraud – with bits of a medieval human cranium married to fragments of an orangutan's jaw. By that time, however, Dawson was long dead, and the fraudster's trail had gone cold.

Could it be that the toad-in-the-hole was a dry run for the Piltdown fraud? Not everyone agrees that Dawson was the culprit. In a frenzy of speculation, nearly every scholar in the vicinity has been accused at some time. The smart money, though, is still on Dawson. Now it looks as though he used the toad-in-the-hole to perfect a style of disclosure that would convince, for a time, the world's leading palaeontologists.

Consider the curious parallels between the two finds. In both instances, Dawson said the objects had been discovered several years earlier. The flint tomb, he told the Brighton naturalists, had turned up 'about two summers ago'. He gave the skull an even longer trajectory, dating finds to 1908 and 1911; then he waited until February 1912 to write to Arthur Smith Woodward, keeper of geology at the British Museum (Natural History) – now London's Natural History Museum.

Both times, too, the reputed finders were local workmen, whom Dawson named, confident that in the class-ridden society of his day no one would ever bother to interview them.

In both cases, Dawson provides exquisite detail. Workmen digging gravel for road repairs found the pieces of skull at Barkham Manor near Piltdown, while the toad-in-the-hole had come from a quarry at the foot of the downs, north-east of Lewes. Its peculiar lemon shape and its comparative lightness had attracted the attention of the men, Dawson claimed. Curious, one labourer, Mr Thomas Nye, broke it open, and found the 'mummied' toad.

In retrospect, the whole thing seems suspicious. The downs are littered with hollow flints, formed around Cretaceous sponges, so Dawson's specimen would hardly be a novelty for the road menders. Equally implausibly, in the Piltdown story he tells us the workmen thought they'd found a coconut, and so decided to smash it. This yarn explained the fragmentary remains, which made disguising the true nature of the fake far easier.

With both hoaxes, Dawson's next move was to associate the specimen with someone in authority who was already a friend. For the flint, he brought in the respected businessman Henry Willett, known for his collection of Cretaceous fossils from the chalk quarries of Lewes. For his later scam, Dawson aimed higher: to validate the skull forgery, he recruited a distinguished colleague he knew from the Geological Society, Woodward of the British Museum. In 1913, Dawson was first author on their joint paper, 'On the Discovery of a Palaeolithic Human Skull', published in the society's journal. Now, at last, Dawson must have imagined, a fellowship of the Royal Society could not be far away.

Alas for Dawson, that ultimate accolade never came. He died suddenly of septicaemia in 1916, aged fifty-two. But he lived long enough to see his skull the talk of the town, and his toad one of the most popular exhibits in Brighton.

The toad's caretaker today, geology curator John Cooper of the Booth Museum of Natural History, is amused by the toad's enduring popularity. At first, he took it at face value, but was troubled by the fact that the amphibian looked considerably bigger in the original 1901 photograph than it does now. For it to have continued to shrink over the ensuing century, it must have only just begun to dry out in 1901. It all fits, Cooper argues. If you had this great idea – you'd seen just the right flint, and were going to concoct a hoax – you wouldn't spend ten years drying a toad. You'd get on with it.

And just how plausible is Dawson's story anyway? Cooper began to wonder. Toads don't hang about on the dry chalk, and if the cobble had at some stage been transported to a wet stream bed, why didn't the entombed toad just rot once it died? Every fossil owes its existence to a series of improbable events, but even so, this toad was pushing it.

So how did Dawson mummify the toad? Did he inject it with alcohol, or dry it in an oven? Perhaps he pickled it. It would be interesting to test it for salt, suggests Cooper, and fitting too, for it was chemical analysis that ultimately revealed the Piltdown skull's true history. Whatever happens, the toad-in-the-hole postcards in the Booth museum shop are likely to remain bestsellers.

Has a fish ever been struck by lightning?

My young neighbour asked me what happens when lightning strikes water. Do all the fish die and what happens to the occupants of metal-hulled boats?

Chris Cooper
Kempston, Bedfordshire, UK

When a bolt of electricity, such as a lightning bolt, hits a watery surface, the electricity can run to earth in a myriad of directions.

Because of this, electricity is conducted away over a hemispheroid shape which rapidly diffuses any frying power possessed by the original bolt. Obviously, if a fish was directly hit by lightning, or close to the impact spot, it could be killed or injured.

However, a bolt has a temperature of several thousand degrees and could easily vaporise the water surrounding the impact point. This would create a subsurface shockwave that could rearrange the anatomy of a fish or deafen human divers over a far wider range – tens of metres.

If someone in a metal-hulled boat was close enough to feel the first effect they would be severely buffeted by the second. Besides which, metal hulls conduct electricity far better than water, so a lightning bolt would travel through the ship in preference to the water.

Andrew Healy
Ashford, Middlesex, UK

When lightning strikes, the best place to be is inside a conductor, such as a metal-hulled boat, or under the sea (assuming you are a fish).

Last century, the physicist Michael Faraday showed that there is no electric field within a conductor. He demonstrated this by climbing into a mesh cage and then striking artificial lightning all over it. Everybody except Faraday was surprised when he climbed out of the cage unhurt.

Eric Gillies
University of Glasgow, UK

Fish don't fart, why is this?

Christine Kaliwoski
Brentwood, California, USA

The writer probably thinks that fish don't fart because she has not seen a string of bubbles issuing from a fish's vent.

However, fish do develop gas in the gut, and this is expelled through the vent, much like that of most animals.

The difference is in the packaging. Fish package their excreta into a thin gelatinous tube before disposal. This includes any gas that has formed or been carried through digestion. The net result is a faecal tube that either sinks or floats, but as many fish practise coprophagia, these tubes tend not to hang around for too long.

Derek Smith
Long Sutton, Lincolnshire, UK

I have on several occasions witnessed my cichlids passing wind to the displeasure of my eel.

This seems to be a result of them taking in too much air while wolfing down flaked foods floating on the surface of the water. If the air was not expelled it would seriously affect their balance.

Peter Henson
University of London, UK

Most sharks rely on the high-density lipid squalene to provide them with buoyancy, but the sand tiger shark (*Eugomphodus taurus*), has mastered the technique of farting as an extra buoyancy device. The shark swims to the surface and gulp air, swallowing it into its stomach. It can then fart out the required amount of air to maintain its position at a certain depth.

Alexandra Osman
London, UK

How did breathing holes evolve in whales and dolphins?

I have always been fascinated by evolution, and while I can usually see why and how certain characteristics evolved in different species, I'm confused by whales and dolphins. How did their breathing holes evolve, bearing in mind their ancestors were land mammals?

Joe Bilsborough
Tarbock, Merseyside, UK

Blowholes are paired nostrils that evolution has shortened and redirected towards the most convenient spot for snorkelling – the top of the head. They do not pass through the brain, though.

As in most swimming, air-breathing vertebrates – such as frogs, crocodiles, capybaras or hippos – whales' nasal openings, or nares, are placed high up so they can breathe with as little raising of the head or snout as possible. They also have protective valves to keep water out.

However, most of the creatures in that list are oriented largely towards the world above: they periodically leave the water for terrestrial activities and they float with nostrils and eyes just above the surface, watching for food and threats.

In contrast, the terrestrial ancestors of ichthyosaurs,

cetaceans and sirenians (manatees and dugongs) evolved into creatures with their attention directed towards the underwater world. Their ears and eyes did not migrate upwards, only their nostrils shortened and the nares migrated towards the highest part of the head because although food and threats no longer came from above, the air they needed to breathe still did. In sirenians that migration is incomplete, so watch this space for another 10 million years.

Antony David
London, UK

The benefit of the location of cetaceans' blowholes is clear, but it's not so clear what factors motivated the initial steps in the migration of the nostrils from the nose to the top of the head. Natural selection certainly does not seem to have made significant progress until the whale's distant ancestors had irreversibly abandoned the land and shifted to a marine lifestyle.

The earliest identified precursor of modern cetaceans is Pakicetus, which lived during the early Eocene, about fifty-three million years ago. It resembled a hyena with hooves, was quite definitely a terrestrial animal and had nostrils at the extreme front of its long snout. It was not until the late Eocene – about twenty million years later – that the first 'true' whale appears in the fossil record. Named *Basilosaurus*, it featured nostrils that had migrated up its snout to a point just in front of its eyes. *Basilosaurus* had nostrils not only shifted backwards in comparison with its forebears, but also converging towards a location on top of the skull, in a clear move towards the modern arrangement. Because *Basilosaurus* was fully aquatic, it seems clear that it was the benefits of the modern set-up that were the driving force behind this particular aspect of its evolution.

Certainly, by the mid-Miocene, some fifteen million years ago, the first modern whales and early dolphins all sported blowholes precisely where they are found on present-day species, although even today evolution has not arrived at a definitive form for a whale's nostrils. Baleen whales, such as the humpback and the blue, have two, while toothed species such as the sperm whale have just one.

Mystery still surrounds the reptilian predecessors in the cetacean's ecological niche. The dolphin-like ichthyosaurs were, as their classical name 'fish-lizard' suggests, particularly well developed for a marine lifestyle. They survived for 140 million years, almost three times as long as whales and dolphins have had to evolve from their terrestrial ancestors.

Yet even the very largest of ichthyosaurs retained two conventional nostrils set just in front of their eyes, very similar to Basilosaurus, despite this configuration requiring them to lift most of their heads out of the water to breathe, and exposing them to attack by predators.

So perhaps the real question is not why whales and dolphins have evolved their manifestly beneficial breathing arrangement, but why their reptilian analogues – and other marine mammals such as the dugong – did not do likewise.

Hadrian Jeffs
Norwich, Norfolk, UK

Can a blindfolded fish still change colour?

How do certain animals, such as the flounder fish, change their colour to match their background? More specifically, if you made a tiny blindfold for the flounder, would it still be able to match its surroundings?

Nick Axworthy
By email, no address supplied

Many fish in the teleost group, such as the minnow, change colour in response to the overall reflectivity of their background. Light reaching their retina from above is compared in the brain to that reflected from the background below.

The interpretation is transmitted to the skin pigment cells via adrenergic nerves, which control pigment movement. Teleost skin contains pigment cells of different colours: *melanophores* (black), *erythrophores* (red), *xanthophores* (yellow) and *iridiophores* (iridescent) Pigment granules disperse through the cell from the centre. The area covered by the pigment at any time determines that cell's contribution to the skin tone.

Many flatfish, including flounder, go further than overall reflectivity and develop skin patterns according to the light and dark divisions of their background. This seems to involve a more discriminating visual interpretation and produces distinct areas of skin with predominantly, but not exclusively, one type of pigment cell. For example, black patches contain mainly melanophores and light patches mainly iridiophores, which can even produce a chequerboard appearance if the fish is lying over a chequered surface.

Since these responses are visual, blindfolding the fish would result in all the components of the chromatic system being stimulated equally. The fish would adopt an intermediate dark or grey skin tone similar to that on a dark night. Over time, hormonal responses via direct light stimulation of the pineal gland through the skull also affect the amount of pigment and number of cells, hence the 'black' plaice sometimes sold in the UK, which have come from the sea around the dark volcanic seabed off Iceland.

Cliff Collis
London, UK

Many animals change the shade or even colour of their skin in response to certain stimuli. In cephalopods such as the cuttlefish, pigment-filled sacs can become extended (flattened) by the action of radially arranged muscle fibres that are controlled by the nervous system. Colour change in these animals is both rapid and spectacular.

In crustaceans and many fish, amphibians and reptiles, specialised dermal pigment-storing cells called chromatophores relocate the pigment internally. The pigment in these chromatophores is either concentrated in the centre of the cell, or dispersed when the pigment fills the cell to the edge.

Imagine a white floor with a small pot of black paint standing in the middle. From above, the floor will look very light, despite a substantial amount of pigment seen as a small black spot in the middle. When the same paint is spread over the floor, the floor looks black. The beautiful trick of the black chromatophores (known as melanophores) is that they can reverse the process, concentrating the pigment in a small area.

Flatfish, such as plaice, flounder and others, are expert at imitating not only the general shade of the surface on which they rest, but also patterns of dark and light material. Not surprisingly, perhaps, their eyes are used to perceive the shade and patterns. Light hitting the retina from above affects the ventral or lower area of the retina, while light reflected from the bottom strikes the dorsal or upper retinal surface.

If the light intensities from the two areas are similar, a signal causes the pigment of the melanophores to be concentrated in the centre of the cell, so the fish turns pale. On the other hand, when the bottom is dark the two areas of the retina receive very different light intensities, and the reverse of the signal causes pigment dispersion and a dark fish. The masters of disguise, the flatfish, can also discern patterns in the bottom surface and imitate them by regulating nerve activity to groups of melanophores.

Stefan Nilsson
Gothenburg University, Sweden

Why are anemones so vivid and varied?

*For a scuba diver, one of the best underwater sights is a rock face covered in brilliantly coloured jewel anemones (*Corynactis viridis*). They exist in many colours,* *and often vivid contrasting colours are found side by side. There are also subdued, semi-transparent variants. Most species of wild animals have evolved to just one or a narrow range of colours, while flowers can have a range of vivid colours, presumably to attract a variety of insects. As far as I know, the anemones aren't trying to attract their prey – it just arrives on the current. So why are they so vivid and so varied?*

George Gall
Truro, Cornwall, UK

My colleague Anya Salih and I have worked on this question for some time in corals, which are close relatives of sea anemones. We believe that the pigments have a protective function against excess light, as discussed in our paper 'Fluorescent pigments in corals are photoprotective', which appeared in *Nature*.

Unpigmented as well as pigmented versions exist in both corals and anemones. The explanation for this is probably that the production of pigments is 'costly', and pigmented versions cope by being fitter than their unpigmented cousins. When conditions are unfavourable the coloured ones do better, though favourable and poor conditions are both common enough that neither form takes over.

Some controversy over this interpretation remains, although we are still waiting for someone to come up with something better.

Guy Cox
Electron Microscope Unit,
University of Sydney, Australia

Will humans ever be able to speak or even understand dolphin?

Riccardo Pesci
Rome, Italy

The idea that dolphins possess a communication system as sophisticated as human language was proposed by John Lilly in the 1960s. Despite loud protests from a sceptical scientific community, Lilly vowed that pioneering researchers would one day 'crack the dolphin code' and begin an interspecies dialogue.

In the ensuing years, dolphins were taught to use artificial symbol systems, with equivocal results. Their performance is comparable to great apes where comprehension is concerned, but when it comes to using symbols to establish two-way communication with humans, dolphins have been overshadowed by linguistic prodigies such as Kanzi the bonobo.

Dolphins' own communication system has been the subject of much study, revealing a perplexing array of vocal and non-vocal signals. But a sober view of half a century's worth of evidence suggests that dolphin communication – even when taking into account the referential, or word-like, nature of their mysterious 'signature whistle' – is nothing more than a variation on the type of communication system seen throughout the animal world. It is complex, to be sure, though likely to be short on content. There is little to suggest that the cacophony of whistles and buzzes is used to share limitless, abstract information in a language-like fashion.

Science is destined to make great strides in unravelling the mysteries of dolphin communication, as there is much we do not yet understand about the function of their vocalisations. However, the idea that dolphins are harbouring a secret language that awaits decryption is looking increasingly like a spot of wishful thinking from a bygone era.

Justin Gregg
Research Associate and Vice-President,
Dolphin Communication Project
Old Mystic, Connecticut, USA

Dolphins and humans can communicate, but is it possible for them to engage in meaningful conversation? Perhaps, but communication between the two species has been limited to date. In fact, there is no compelling scientific evidence that humans and dolphins can engage in exchanges of information beyond those that involve a human requesting a dolphin to perform some behaviour or those that inform a human about some object the dolphin would like but cannot obtain without human assistance.

There are many possible reasons why we cannot converse with dolphins, including the fact that we have much to learn about their communication systems. Dolphins produce a variety of sounds and other behavioural cues that appear to be communicatively significant, yet the communication units used by dolphins remain unknown. For example, are whistles separate single units or some combination of smaller units? Once the units that comprise the dolphin communication system are ascertained, the daunting task of determining what they mean remains. This will require a comparison of how individual units are used in isolation and with other units in a variety of contexts, a process that has only just begun.

Clearly, such work is necessary before conversations can occur. However, it may not be sufficient. Conversations require shared interest in a topic, so humans will need to find subjects that interest dolphins. Given the differences between us, discovering a common ground for meaningful conversations may be more difficult than some humans imagine.

Stan Kuczaj
Marine Mammal Behavior and Cognition Laboratory
University of Southern Mississippi

Why do flying fish fly?

Why do flying fish fly? Is it to escape predators, or to catch flying insects, or as a more efficient means of getting around than swimming? Is there some other entirely different reason?

Julyan Cartwright
Palma de Mallorca, Spain

The usual explanation for flight in flying fish is as a way to escape predation, particularly from fast-swimming dolphin-fish. They do not fly to catch insects; flying fish are largely oceanic and flying insects are rare over the open sea.

It has been suggested that their flights (which are actually glides, because flying fish do not flap their 'wings') are energy-saving, but this is very unlikely as the vigorous take-offs are produced by white, anaerobic muscle beating the tail at a rate of 50–70 beats per second, and this must be very expensive in terms of energy use.

Flying fish have corneas with flat facets, so they can see in both air and water. There is some evidence to suggest that they can choose landing sites. This might allow them to fly from food-poor to food-rich areas, but convincing evidence of this is lacking.

There seems to be little doubt that escape from predators is the major purpose of their flight, and this is why so many fly away from ships and boats, which they perceive to be threatening.

John Davenport
University Marine Biological Station
Millport, Ayrshire, UK

Strictly speaking, the flying fish does not fly, it indulges in a form of powered gliding, using its tail fins to propel it clear of the water. It sustains its leap with high-speed flapping of its oversized pectoral fins for distances of up to 100 metres. The sole purpose of this activity seems to be to escape predators. If one can manage to tear one's eyes away from the magic of the unexpected and iridescent appearance of a flying fish, a somewhat more substantial fish can often be seen following its flight path just below the surface.

Tim Hart
La Gomera, Canary Islands, Spain

I have seen whole schools of flying fish become airborne as they try to escape tuna which are hunting them, and minutes later have seen the school of tuna attempt similar aerobatics as dolphins move in for their supper of tuna steaks.

A morning stroll around the decks of an ocean-going yacht will often provide a frying pan full of flying fish for breakfast. Presumably they are instinctively trying to leap over a predator (in this case the boat) but as they don't seem to be able to see too well at night they land on the deck. They rarely land on board during the day. Most alarmingly they will land in the cockpit, and even hit the stargazing helmsman on the side of the head.

Don Smith
Cambridge, UK

Creature Curiosities

Fish farting may not just be hot air

In 2003, biologists linked a mysterious underwater farting sound to bubbles coming out of a herring's anus. No fish had been known before to emit sound from its anus or to be capable of producing such a high-pitched noise. 'It sounds just like a high-pitched raspberry,' said Ben Wilson of the University of British Columbia in Vancouver, Canada. Wilson and his colleagues could not be sure why herring made this sound, but initial research suggested that it might explain the puzzle of how shoals keep together after dark.

'Surprising and interesting' was how aquatic acoustic specialist Dennis Higgs, of the University of Windsor in Ontario, described the discovery. It was the first case of a fish potentially using high frequency for communication, he believed. Arthur Popper, an aquatic bio-acoustic specialist at the University of Maryland, was also intrigued. 'I'd not have thought of it, but fish do very strange and diverse things,' he said.

Fish are known to call out to potential mates with low 'grunts and buzzes', produced by wobbling a balloon of air called the swim bladder located in the abdomen. The swim bladder inflates and deflates to adjust the fish's buoyancy.

The biologists initially assumed that the swim bladder was also producing the high-pitched sound they had detected. But then they noticed that a stream of bubbles expelled from the fish's anus corresponded exactly with the timing of the noise. So a more likely cause was air escaping from the swim bladder through the anus.

It was at this point that the team named the noise Fast Repetitive Tick (FRT). But Wilson pointed out that, unlike a human fart, the sounds were probably not caused by digestive gases because the number of sounds did not change when the fish were fed. The researchers also tested whether the fish were farting from fear, perhaps to sound an alarm.

But when they exposed fish to a shark scent, there was again no change in the number of FRTs. Three observations persuaded the researchers that the FRT was most likely produced for communication. Firstly, when more herring were in a tank, the researchers recorded more FRTs per fish. Secondly, the herring were only noisy after dark, indicating that the sounds might allow the fish to locate one another when they could not be seen. Thirdly, the biologists knew that herring could hear sounds of this frequency, while most fish cannot. This would allow them to communicate by FRT without alerting predators to their presence.

Wilson emphasised that this idea was just a theory. But the discovery is still useful, he said. Herring might one day be tracked by their FRTs, in the same way that whales and dolphins are monitored by their high-pitched squeals.

MAMMALIA

FROM BATS TO ELEPHANTS, AND EVERYTHING IN BETWEEN

Is it true that elephants are the only quadrupeds that cannot jump?

Tad and Lydia Forty (aged 13 and 8)
Bath, Somerset, UK

This is a fun question, but it is not true even if we include only four-legged animals that routinely walk on land.

Elephants cannot jump, from level ground anyway. This is true even when they are babies, as far as we know, but they are not alone. Probably all turtles cannot truly jump. It is also likely to be true for some salamanders and large crocodiles, some chameleons and other lizards. In fact, the statement is almost certainly not true even if restricted to mammals. Hippos probably cannot or do not jump, along with moles and other burrowing mammals, sloths, slow loris and other climbing specialists.

However, the truth is that no researchers have looked at this question in a rigorous way. We don't even know specifically why – in terms of detailed anatomical mechanisms and physics – any of these animals cannot jump. There are just scattered anecdotes and folklore, like the tired myth that elephants have four knees, which I still encounter again and again from the public. Elephants actually have two knees like all other mammals because their anatomy is essentially the same.

So the question is certainly worth addressing. But there are a lot of species out there, so as a general rule it's probably best to assume there is unlikely to be any species that is alone in being unable to do some seemingly common activity.

John R. Hutchinson
Reader in Evolutionary Biomechanics
Royal Veterinary College
University of London, UK

Racehorses weighing about half a tonne are among the largest quadrupeds that can make impressive jumps. In horse racing, the Chair, the highest fence on the Grand National course, is 1.8 metres high.

The largest wild animal I have seen making an impressive jump was an eland, one of a group that I saw galloping in Kenya. Its jump was high enough to have cleared the back of another eland, roughly 1.4 metres from the ground. The animal probably weighed about the same as a racehorse.

Large male African elephants weigh around 5 tonnes, and Asian elephants only a little less. After them, the heaviest quadrupeds are the hippopotamus (about three tonnes) and the white and Indian rhinos (about 2 tonnes). Whether these and other large animals can jump depends on what you count as jumping. A film I took of a white rhino galloping at 7.5 metres per second showed that, at one stage of its stride, all four feet were off the ground. I do not think of that as jumping, but I cannot think of any clear-cut definition of jumping that would exclude it.

Big jumps require strong leg bones and muscles. The

vertical component of the force the feet exert on the ground, averaged over a complete stride or jump, must equal the animal's weight. In a substantial jump, the animal is off the ground for longer than it would be in a running stride, so its legs will be subject to larger forces at take-off and landing.

Simple physics tells us that if big animals were precisely scaled-up versions of smaller ones, their weights would be proportional to the cubes of their linear dimensions. The cross-sectional areas of bones and muscles, however, would be proportional only to the squares. An animal with double the linear dimensions of another would be eight times as heavy, but its legs would be only four times as strong, and so less able to jump.

Of course, even closely related animals of different sizes are not scale models of each other. For example, a 500-kilogram eland has relatively thicker, straighter legs than a 5-kilogram dik-dik – but the differences are not sufficient to eliminate the disadvantage for large jumpers.

Other than size, a quadruped's anatomy or physiology may be unsuitable for jumping. Some desert lizards that burrow in loose sand have greatly reduced limbs, tortoises have very slow muscles and the limbs of moles are highly modified for digging. I have never seen any of those quadrupeds jump, and do not expect to.

R. McNeill Alexander
Emeritus Professor of Zoology
University of Leeds, UK

Elephants are not the only quadrupeds that cannot jump. Some of the quadruped dinosaurs could not jump, including apatosaurus and diplodocus.

Edward Rivers (aged 7)
Angmering, West Sussex, UK

Really heavy animals like rhinos and hippos can hardly jump or land without injury. After reaching terminal velocity, mice would bounce after hitting the ground whereas elephants would break, or, according to urban legend, 'splash'.

Even so, don't jump to optimistic conclusions if a large animal chases you over a ditch. Does it still count as 'being able to jump' if the jump causes the animal injury? If so, then you are in trouble because, yes, Indian elephants can jump. J. H. Williams in his book *Elephant Bill* relates how a stampeding female jumped a ditch handily, though she went lame in both forefeet soon after.

Jon Richfield
Somerset West, Western Cape, South Africa

Why did the sheep run up the road?

Why do sheep always run in a straight line in front of a car and not to the side?

Aled Wynne-Jones
Cambridge, UK

Sheep and other animals run ahead of cars because they do not realise that cars cannot climb grassy banks. Ancestral sheep were pursued by wolves and big cats. If an animal tries to turn aside some yards from the hunter, the pursuing animal sees what is happening, makes an easy change of course and intercepts the victim, which is presenting its vulnerable flank.

If, however, the prey dodges at the last minute, the outcome is different. The hare is the master of this strategy: as the greyhound reaches out with its jaws, the hare jinks to one side and the dog overshoots or, with luck, tumbles head over heels.

The instinctive response of a sheep or a hare to an approaching car is at least not as maladaptive as that of the hedgehog.

Christine Warman
Clitheroe, Lancashire, UK

Herbivores are killed by predators who normally grab them by the throat while running alongside, so the prey will always do its best to keep a potential threat behind its tail, swerving as the predator attempts to overtake. That's why a kangaroo, seeing a car drawing alongside, will jump onto the road right ahead in order to keep the car directly behind it, and often get run over in the process. As long as a car proceeds in a straight line behind a sheep, the sheep will try to outrun it in a straight line.

G. Carsaniga
Sydney, Australia

Sheep are much underrated. They don't merely run in a straight line – they run straight for a while, then dive to the side. This is not woolly thinking, it's perfectly logical.

Sheep loose in the road are usually confined to country areas, where roads are bounded by steep verges, cliffs, hedges, fences and ditches. The sheep recognises that if it cannot beat the car on the flat, it stands no chance whatsoever up a bank, so it attempts to outrun the vehicle down the road.

What happens then is that the vehicle slows, and when it reaches a speed that is slow enough for the sheep to think it might beat the car over the obstructions at the side of the road, it swerves. And since most of the time this action is proved correct (most vehicles don't follow sheep off the road), the sheep carries on behaving in this way. QED, by sheep logic.

Clearly, this is a much more successful approach to road safety than that shown by humans. Humans rarely try to outpace the oncoming car. They tend to dive to the side of the road. Since more people are run over than sheep, one can conclude we have much to learn from sheep logic.

William Pope
Towcester, Northamptonshire, UK

Sheep, being clever animals with an instinctive grasp of psychology, know that most drivers, though enjoying an occasional kill as long as they can use the 'it jumped in front of me, there was nothing I could do' excuse, are not so depraved as to deliberately run something down. Thus running in a straight line has a distinct advantage over veering to the side.

Erik Decker
National Institute of Animal Husbandry
Department of Cattle and Sheep
Tjele, Denmark

Why aren't there more green mammals?

The benefits of camouflage would suggest that there should be green mammals. Are there any – and if not, why not?

A. C. Henderson
Braco, Perth and Kinross, UK

There is only one green mammal, the three-toed sloth. This is because a coat of algae covers the sloth's fur. Because of the sloth's tardiness and lack of personal hygiene, this is never cleaned off. No known mammal is capable of producing its own green epidermal pigment. The main reason for the absence of green mammals seems to be an ecological one. In general, mammals are simply too big to use a single colour for camouflage as there are no blocks of green large enough to conceal them. Most mammals have an environment that is made up of patches of light and dark and composed of many different colours. This means that those mammals which are camouflaged tend to be dappled or striped. Animals that do use green coloration for camouflage, such as frogs and lizards, are small enough to use solid blocks of green – leaves and foliage – for cover.

Paul Barrett
Department of Earth Sciences
University of Cambridge, UK

The main predators of most mammals are other mammals, especially the carnivores, such as the cat, dog and weasel families. Carnivores are all colour-blind or, at best, have very limited colour vision. Hence effective camouflage against them is not a matter of coloration but of a combination of factors such as brightness, texture, pattern and movement.

Graeme Ruxton
Scottish Agricultural Statistics Service
Edinburgh, UK

Your answers to this question only mention the tree sloth, which is not truly green, just covered with algae. There is actually a real green mammal – the green ringtail possum (*Pseudocheirus archeri*) – and what a lovely animal it is. The possum is a marsupial endemic to a small area in north-eastern Australia. You can see a fine colour portrait of it in *The Complete Book of Australian Mammals* (Angus and Robertson, 1983). The article accompanying the portrait is by J. W. Winter, an expert on the possums of that part of Australia. He writes: 'This remarkably beautiful ringtail is aptly named: a mixture of black, grey, yellow and white hairs confers a most unusual lime-green colour to its thick, soft fur.'

I would add that it is also the most docile wild animal I have ever encountered. A scientist studying possums during the 1960s who caught one and kept it for a day before returning it to the wild allowed me to photograph it. It made no attempt to struggle, scratch or bite when taken out of the cage, nor did it try to escape.

Winter makes no comment as to whether the green

colour has any apparent advantages, but he does report that 'its daytime roost, unlike that of other possums, is usually on an open branch. It sleeps upright, curled into a tight ball, gripping the branch with one or both hind feet and sitting on the base of its coiled tail, with the forefeet, face and tip of the tail tucked tightly into its belly.'

A motionless, amorphous green ball among the multitudinous shades of green in the rainforest would be far from obvious. The only predators Winter reports (apart from Aboriginal humans in the past) are nocturnal: the rufous owl (*Ninox rufa*) and the spotted-tailed quoll (*Dasyurus maculatus*). The latter is a marsupial carnivore, with a head and body length of about half a metre, found over much of eastern Australia including Tasmania.

H. S. Curtis
By email, no address supplied

Do animals need glasses?

I would estimate that about 40 per cent of people that I know need glasses or contact lenses for distance vision. Assuming that this sample is typical of the human race, I would like to know why it is that eye problems prevalent in humans such as myopia (short-sightedness) seem very rare in wild animals. As far as I know, myopia is a genetic condition and so is not usually acquired by habits such as reading small print (otherwise one would expect recovery after stopping the habit). Obviously, it is not easy to test the eyesight of an animal, but if the incidence of myopia is as high in wild animals as it is in humans then how can the animals survive?

Stephen White
Surbiton, Surrey, UK

For most nonhuman mammals the ocular refraction (the lens power required to form a clear retinal image of an object at infinity) for optimal distance vision tends strongly towards emmetropia, or perfect vision. Similarly, the variation of distance refraction and the presence of astigmatism is lower than for humans. It should be noted, however, that the sample tested is far smaller for mammals than for humans.

These trends have been noted by my co-workers Clive Phillips, Jacob Sivak, Robin Best and Bill Muntz, and myself, in a number of different studies using standard clinical optometric techniques (obviously avoiding those that require a verbal response).

The studies have covered such disparate species as domestic sheep, guanaco (a llama), polar bear, manatee and three-toed sloth. Such findings would support the suggestion that a myopic mammal would be at a natural disadvantage.

David Piggins
Bangor, Gwynned, UK

There probably is a genetic factor in short-sightedness, but that does not explain why it is so common in modern society. Those who regularly focus their eyes over longer distances, such as sailors and mountaineers, are apparently less likely to become myopic. It seems likely that the muscles on either side of the eye can be trained to contract the eye, thus overcoming short-sightedness. Once a person starts wearing glasses, the need for such adjustment disappears.

Brynjolfur Thorvardarson
Southampton, Hampshire, UK

The fact that about 40 per cent of people you know wear glasses or contact lenses does not really indicate a malfunction in the entire human race. If you travel among primitive peoples of the world, you will find numerous examples of keen sight that seem almost super-human to the Western mind.

Kevin Wooding
Oxford, UK

Myopia often has a genetic component, but this isn't the whole story. People who do close work are often myopic (tailors are the classic example) but in the past it was usually assumed that it was their myopia that attracted them to such jobs: the idea that myopia could be acquired seemed too far-fetched. However, several decades ago it was observed that university students with better than average grades (who presumably read more) tended to be myopic, as were laboratory animals that were raised in a confined environment.

C. R. Cavonius
University of Dortmund, Germany

Primates brought up in captivity do tend to become myopic. Myopia is caused by the axial length of the eye, but changes in corneal power also affect sight. Most change is likely to occur in the growth phase during the few years after birth, but may continue to a lesser extent for the next three decades in humans.

Better evidence comes from chicks, where ten dioptres of myopia or hypermetropia can be induced by contact lenses, and reversed, over a few weeks. There is a dramatic change in the posterior segment of the eye, which is accounted for by alterations in the rate of growth of the eye as the chick ages. Just like humans who squint to improve their vision, chicks have the capacity for corneal and lenticular accommodation, but this appears to exert little influence upon growth.

Whether human myopia can be arrested or reversed is the subject of some debate in ophthalmology. There does seem to be something approaching an epidemic of myopia, especially in the Far East, which cannot be explained purely by genetic or occupation selection.

Matt Cooper
Brighton, East Sussex, UK

Among older people, acquired defects of vision are more common than inherited myopia. At ages beyond those that would have been achieved in the wild, human eye tissues distort and lose flexibility, causing astigmatism and presbyopia. The lens may go opaque from cataracts, or it may darken so that it needs more light.

Domestic animals may show similar defects when they get beyond the life expectancy in the wild, and old dogs or cats often go blind from cataracts. Replacing a pet dog's lens is practically a routine operation nowadays. One need not give the dog glasses afterwards – just as long as it can see clearly enough to get at its food dish. It doesn't have to read the brand name on the tin.

Jon Richfield
Somerset West, Western Cape, South Africa

Does cows' milk taste different in summer?

Cows eat lush grass in summer but prepared dried feed in winter. Does the milk that I pour on my breakfast cereal differ in any way throughout the year?

Graeme Mawson
Newcastle upon Tyne, UK

More than fifty years ago, I spent some time living on my aunt's croft in Sutherland.

In winter, her cow Bella was fed largely on turnips, and during those months Bella's milk also tasted of turnip. But this taste disappeared in summer when she was fed on grass.

Nowadays, dairy cows are fed on a fermented stored grass called silage during the winter and this is supplemented with processed feed containing animal protein.

The winter milk no longer smells or tastes of turnip,

but I assume that if turnip used to contribute to the composition of milk then the modern winter feed probably does too. In retrospect, I think I would prefer the turnip.

Ian Sutherland
Birmingham, UK

Your correspondent probably assumes that the milk he pours onto his breakfast cereal every morning is much the same as the milk that comes out of a cow. In fact, most drinking milk is pasteurised and standardised for fat, so this quality parameter is invariable throughout the whole year. Milk may in future also be standardised for protein – in effect removing high-value solids from the milk for further processing.

As a dairy farmer drinking raw milk, both I and my family might detect slightly higher levels of fat and lactose at certain times of the year but, unless the cows graze on a patch of wild garlic, the taste of the milk remains the same.

I believe that in parts of France, certain cheeses are only made from milk produced in a particular season and from cows that graze on pastures with specific herbs.

Pasteurising milk probably removes taste by damaging natural enzymes. If anyone could achieve the antibacterial effects of pasteurisation without also damaging the taste, the process would be invaluable to the dairy industry and would also enable your correspondent to experience the taste of milk straight from the cow.

Mark Pearse
South Molton, Devon, UK

The effect of different feeds on the content of milk is actually only slight. The nutrients affected are the fat-soluble vitamins A and E, folate and iodine. There is no significant difference in any of the macronutrients such as protein, carbohydrate and fat.

Vitamin A in whole milk varies from 69 micrograms per 100 millilitres in the summer to 44 micrograms in the winter. Whole milk is a useful source of vitamin A, particularly for young children, and this is one of the main reasons why milk is recommended for children under two years.

The variations in vitamin E and folate are both small: vitamin E averages 0.1 milligrams per 100 millilitres in summer and 0.07 milligrams in winter, while folate rises from 4 micrograms per 100 millilitres in summer to 7 micrograms in winter.

Iodine content averages 7 micrograms per 100 millilitres in summer and 38 micrograms in winter because cattle consume greater amounts of iodine-supplemented manufactured feed in the winter months.

The widespread addition of iodine to animal feed has meant that milk and dairy products are now major sources of iodine in the British diet – a factor which has helped significantly to eliminate goitre.

Sarah Marshall
National Dairy Council, London, UK

Why is red meat red and white meat white?

What is the difference between the various animals that makes their flesh differently coloured?

Tom Whiteley
Bath, Somerset, UK

Red meat is red because the muscle fibres which make up the bulk of the meat contain a high content of myoglobin and mitochondria, which are coloured red. Myoglobin, a protein similar to haemoglobin in red blood cells, acts as a store for oxygen within the muscle fibres.

Mitochondria are organelles within cells which use oxygen to manufacture the compound ATP which supplies the energy for muscle contraction. The muscle fibres of white meat, by contrast, have a low content of myoglobin and mitochondria.

The difference in colour between the flesh of various animals is determined by the relative proportions of these two basic muscle fibre types. The fibres in red muscle fatigue slowly, whereas the fibres in white muscle fatigue rapidly. An active, fast-swimming fish such as a tuna has a high proportion of fatigue-resistant red muscle in its flesh, whereas a much less-active fish such as the plaice has mostly white muscle.

Trevor Lea
Oxford, UK

The colour of meat is governed by the concentration of myoglobin in the muscle tissue which produces the brown colouring during cooking.

Chickens and turkeys are always assumed to have white meat, but free-range meat from these species (especially from the legs) is brown. This is because birds reared in the open will exercise and become fitter than poultry grown in restrictive cages. The fitter the bird, the greater the ease of muscular respiration, and hence the increased myoglobin levels in the muscle tissue, making the meat browner.

All beef is brown because cattle are allowed to run around in fields all day, but pork is whiter because pigs are lazy.

T. Filtness
Winchester, Hampshire, UK

At the risk of flogging a dead, er, penguin, why don't polar bears' feet freeze?

Paul Newcombe
Zurich, Switzerland

Unlike the penguin with its fancy internal plumbing (see page 144), the reason that polar bears' feet do not freeze is good insulation, pure and simple.

Polar bears (*Ursus maritimus*) are just about the best-insulated animals on the planet, certainly among those species of mammal that do not live primarily immersed in water. An adult bear has ten centimetres of blubber beneath its skin, which in turn is covered by a thick coat of fur.

This fur relies not only on its density, but also on its unique structure: each hair is a hollow tube, so that air is trapped inside the hairs as well as between them. Even without covering its nose with its paws (as it is reputed to do, although the evidence is very limited) a polar bear is almost invisible to heat-sensitive infrared photography or the latest military image-intensification technology.

The polar bear also has very hairy pads on its feet, and the tough skin is extremely callused on the underside of the paws, so there is a sturdy layer of dead tissue between the ice and any blood vessels.

There may also be another factor at work. The underside of a polar bear's paw is dotted with dozens of papillae – small nipple-shaped extrusions of even more callused skin – which provide extra grip in the same way as the studs on a footballer's boot. It is these papillae that enable a polar bear to accelerate to a very respectable pace on the ice and overcome its awesome inertia. They also prevent it skating out of control, past a potential meal.

On really compacted ice, the bears tend to lift part of the underside of the paw clear of the surface. The papillae enable an additional cushion of insulating air to be trapped between most of the pad and the ice.

Such highly developed thermal adaptations can, however, be a double-edged sword. A bear attempting a brisk trot in ambient temperatures of 10°C or greater would succumb, almost immediately, to a fatal attack of heat stroke. During the Arctic summer it can often be far hotter than that, limiting the polar bear's ability to function as a hunter.

This potential cramping of the polar bear's style may prove as fatal to the species' chances of survival as the actual destruction of its territory. If global warming causes the polar bear to die out, it would surely be the most terrible irony that this was because it had mastered the art of conserving the very energy that a profligate humanity has squandered so obscenely.

Hadrian Jeffs
Norwich, Norfolk, UK

Could polar bears and penguins relocate?

If polar bears were transferred to Antarctica could they survive? And would penguins survive in the Arctic?

Richard Davies
Swansea, UK

Polar bears would probably survive in the Antarctic, and the Southern Ocean around it, but they could devastate the native wildlife. In the Arctic polar bears feed mainly on seals, especially young pups born on ice floes or beaches. Many of the differences in breeding habits between Arctic and Antarctic seals can be interpreted as adaptations to evading predation by bears.

Polar bears would find plenty of fish-eating mammals and birds around Antarctica. Penguins would be particularly vulnerable because they are flightless and breed on open ground, with larger species taking months to raise a single chick. Bears can only run in short bursts, but they could catch a fat, sassy penguin chick or grab an egg from an incubating parent.

In the Arctic polar bears hunt mainly on the edge of the sea ice, where it is thick enough to support their weight but thin enough for seals to make breathing holes. The numerous islands off the north coast of Canada, Alaska and north-west Europe provide plenty of suitable habitats. The Antarctic continent is colder, with only a few offshore islands, so bears would probably thrive at lower latitudes in the Southern Ocean than in the Arctic.

We can only hope that nobody ever tries what the questioner suggests. Artificially introduced predators often devastate indigenous wildlife, as it is not accustomed to dealing with them. This occurred with stoats in New Zealand, foxes and cats in Australia, and rats on many isolated islands.

Large, heavy animals would also trample the slow growing, mechanically weak plants and lichens of the Antarctic. For instance, Norwegian reindeer have decimated many native plants in South Georgia, an island in the South Atlantic Ocean, since they were introduced eighty years ago.

C. M. Pond
Department of Biological Sciences
The Open University
Milton Keynes, Buckinghamshire, UK

While, as far as I know, no one has ever been stupid enough to introduce polar bears into the Antarctic, there have been at least two practical attempts to transplant penguins to the Arctic.

The original 'penguin' was in fact the late great auk (*Pinguinus impennis*), once found in vast numbers around northern shores of the Atlantic. Although no relation to southern hemisphere penguins, it was very similar in appearance, and filled much the same ecological niche as penguins, particularly the king penguins of the sub-Antarctic region.

With any attempt to introduce an alien species, there must actually exist an appropriate ecological niche for it to fill, and it must be vacant. For the most part, the ecological niches occupied by penguins in the south are filled by the auk family to the north. But the demise of the great auk in the mid-nineteenth century at the hands of hungry whalers created not only a vacancy that one of the larger penguins might neatly slot into, but also a potential economic demand for the penguin's fatty meat and protein-rich eggs.

It was perhaps the possible economic opportunities that prompted two separate bids to introduce penguins into Norwegian waters in the late 1930s. The first, by Carl Schoyen of the Norwegian Nature Protection Society, released groups of nine king penguins at Røst, Lofoten, Gjesvaer and Finnmark in October 1936. Two years later, the National Federation for the Protection of Nature, in an equally bizarre operation, released several macaroni and jackass penguins in the same areas, even though these smaller birds would clearly find themselves competing directly with auks or other native seabirds.

The outcome was unhappy for the experimenters and, most particularly, for the penguins. Among those whose fate is known, one king was quickly despatched by a local woman who thought it was some kind of demon, while a macaroni died on a fishing line in 1944, although from its condition it had apparently thrived during its six years in alien waters.

And it soon became obvious that the real reason why any attempt to fill the ecological gap left by the great auk was destined to fail was the very reason that the niche was vacant in the first place – such large seabirds could not happily coexist with a large and predatory human population. Of course, it is the steadily increasing human presence in the far south that is now threatening penguins in their native habitat.

Hadrian Jeffs
Norwich, Norfolk, UK

Why don't bats get dizzy when they hang upside down? Or do they?

Year 5, Christopher Hatton School
London, UK

Dizziness is a sensation humans describe when they feel a sense of motion, even when not moving. It can be associated with queasiness or nausea, and sometimes vomiting. Other types of dizziness include motion sickness and vertigo, which often manifests itself as a spinning feeling, or other sensations such as light-headedness or heavy-headedness.

It is impossible to know for sure whether or not an animal is dizzy, because it cannot communicate such feelings. However, it is possible to infer an animal is dizzy from how it behaves. For example, if an animal is aimlessly walking in circles, it is probably dizzy.

Motion sickness occurs when there is excessive stimulation of the inner ear or from a conflict between sensory information from different sources, such as from the inner ear and the eyes. The balance mechanism of the inner ear is complicated, and includes sensors that detect both movement and orientation with respect to gravity, even when an individual is not moving. Bats have such a balance mechanism, and in addition use echolocation.

The parts of the inner ear that are important for orientation with respect to gravity are called the otolith organs: the utricle and the saccule. It is these parts of the inner ear that would be activated while the bat was hanging upside down. Stimulating these parts of the inner ear, however, would not necessarily lead to dizziness, especially in a dark cave where there is no conflict between information from the inner-ear balance mechanism and vision.

The bottom line is that bats are used to hanging upside down without showing any behavioural changes that would suggest dizziness or motion sickness. But because we cannot ask a bat directly whether or not it is dizzy, we cannot be certain about the effects of hanging upside down.

Joe Furman
Editor of the *Journal of Vestibular Research*
University of Pittsburgh
Pennsylvania, USA

When you think of bats, you usually think of them in one of two conditions: hanging upside down resting, or flitting about pulling high-g turns in the dark. So why don't they get dizzy?

Bats have evolved a number of adaptations to allow them to hunt and hang without the problems that humans would face. First, some bats have specialisations in the vestibular portion of their inner ears – the portion that generates sensory signals for controlling balance. Their sacculus, which in humans acts as a gravity sensor to help us stand upright, is slightly rotated forwards. This enables it to act more as a pitch detector, which is more useful in flight. Second, their circular canals, which sense rotation of the head, have an internal structure more like a bird's than a human's. This probably allows them to make high-speed turns without the fluid in the canals sloshing back and forth too much. Lastly, if you photograph bats in flight with a high-speed camera, you notice that they keep their heads very stable except in the most violent turns.

But it is the way in which bats sense the world that probably gives them immunity to dizziness. All the vestibular system does is tell you about changes in acceleration of your head. It requires other senses to pin down your position and motion in the outside world. We primarily use vision to do this, but vision is very slow. Anything you look at that takes a second or less to cross 30 degrees of your vision appears smeared. Echolocating bats, while not blind, rely more on biosonar, an especially precise form of hearing that lets them build up 3D images from echoes.

Echolocating bats emit brief sonar chirps from 30 to more than 150 times per second, and respond to changes in echoes of less than a microsecond. These bats integrate echolocation with their vestibular system, so they are working with a faster, more precise positioning system than humans do with vision. Because dizziness and motion sickness usually arise when signals from the vestibular system conflict with those from other sensing systems, bats are less likely to show motion sickness than other mammals.

Seth Horowitz
Assistant Professor of Neuroscience
Brown University, Rhode Island, USA

Why did American cowboys need to shoe their horses, but the Native Americans did not?

Larry Curley
Huntingdon, Cambridgeshire, UK

The answer to this stems from evolution; not just biological evolution but also variations in technological and social evolution on both sides of the Atlantic.

The horse's native habitat is large grassy plains with generally dry climate, such as the steppes of central Asia, where the wild ass originated, the African savannah, which is home to near-relatives such as the zebra and the now-extinct quagga, and the prairies of North America where the genus *Equus* evolved.

Horses were driven to extinction in North America

about 7,600 years ago, possibly by climate change or hunting by the ancestors of Native Americans. They only returned to the New World when the Spanish brought them there in the sixteenth century.

The tribes of the North American plains and the American Southwest came across these horses, or at least their feral descendants, after they began to escape from Europeans around 1540. Newly established wild herds spread up the Mississippi valley, where most of the tribes had a settled agrarian lifestyle. The Plains Indians led a nomadic existence, and despite never having seen a European – mounted on a horse or otherwise – it was they who realised the horse's potential for enhanced mobility.

First to mount up were the Kiowa and tribes of the Missouri valley, who were riding horses by the 1680s. By 1714 the Comanche of Wyoming had joined them on horseback, followed by the Snake of southern Idaho and eastern Oregon, and the Cheyenne of Minnesota and North Dakota, who in the 1730s introduced horsemanship to their neighbours, the Teton Sioux. Finally, the Sarcee tribe of Canada became the most northerly mounted tribe by 1784.

In the space of just over a century the horse had transformed Native North American society, and not always for the better. The arable society of the Missouri valley was ultimately destroyed by raiding 'war parties' of Comanche, Cheyenne and Dakotas, well before Europeans began their genocidal march.

The Plains Indians were hunter-gatherers, with no significant metal-working skills, so any metal goods were obtained through trade with Europeans and hence were at a premium.

Even if they had needed them, horseshoes would always have been a technology beyond their socio-economic resources. But they had no need to shoe their horses.

The horseshoe was developed to meet the conditions faced by domestic horses in north-west Europe, where, judging from archaeological evidence, they were probably first produced by the Gauls or Franks in the fifth century. Europeans needed horseshoes because of a combination of climate, terrain and pattern of use, with the generally wet weather and soft, heavy soils acting to soften the normally calloused sole of the hoof. Horses were used for travel and in wars. They were often heavily laden while travelling at quite high speeds, which placed great stress on the hooves, often causing them to wear unevenly and eventually split, rendering the animal lame and useless.

The lifestyles of horses used by Plains Indians, on the other hand, differed little from that of their ancestors in the wild. The animals moved together in large numbers at relatively low speeds, over flat, arid steppe country. As a result, their hooves were harder and wore more evenly. In addition, Native American warriors had more than one mount each, with one band of 2,000 Comanche braves keeping a string of 15,000 horses in tow.

The quality of husbandry among European settlers often left a great deal to be desired. The US cavalryman at the time of the Indian wars in the eighteenth and nineteenth centuries was usually an indifferent horseman, and more concerned with his own well-being than that of his mount, which was after all government property.

By contrast a Plains Indian brave's horses were his fortune and livelihood, and he cared for them accordingly if he valued his life. Cowboys riding the range were in a very similar position, which was why horse-stealing was considered the worst of all crimes in the old west, as it was tantamount to murder.

However, the peculiarities of their profession, and the specific qualities demanded from the horse required the use of shoes. The classic cowboy's mount was the quarter horse, the fastest steed in the world, but only over short distances (the quarter-mile that gave it its name). This enabled a cowboy to race from one point around a large herd to another at short notice and in short order, but applied stresses a bare hoof could not sustain in the long term.

Hadrian Jeffs
Norwich, Norfolk, UK

Where are all the three-legged animals?

None of the countless species of animal in existence has three legs. Creatures such as the kangaroo and the meerkat use their tails for balance, but a tail is plainly not the same as a leg. This pattern does not apply only to mammals – other kinds of animal have an even number of legs, too. Why wouldn't having three legs work?

Monika Hofman
London, UK

A tripod is wonderfully stable, so there could be something to be said for having three legs. When insects walk, they use their legs as two sets of three. At any instant their weight is supported by three legs – two on one side of the body and one on the other. Meanwhile, the other three legs can be moved forward to form the next 'tripod'.

All the animals mentioned in the question are bilaterally symmetrical, so it is not surprising that their limbs come in pairs – two in the case of land-dwelling mammals, three in insects, four in spiders, and various larger numbers in crustaceans, centipedes and millipedes.

In contrast, starfish are built on a radially symmetrical plan (also seen in sea urchins and sea cucumbers), so they often have five arms. However, these are not like legs, in that they are not manipulated for locomotion. Starfish move using thousands of hydraulically operated tube feet, arranged along the undersides of their arms.

If you had to walk on exactly three legs – as opposed to the insect's two sets of three – you would not want an asymmetrical gait with two legs on the left and one on the right, or vice versa. But an arrangement with one leg on the midline and one on each side is certainly feasible. Having recently been getting about on one leg and a pair of crutches, I can confirm that you can move quite quickly this way, though it is tiring and more difficult on slopes and steps than using two legs.

I think we have to conclude that three legs is an unlikely arrangement in a bilaterally symmetrical animal, and seems to confer no advantage in movement over two or four.

John Gee
Aberystwyth, Ceredigion, UK

As a long-term user of crutches, I walk with three 'legs' as often as not. Quite a few gaits are possible while your weight is borne by two legs and crutches, but if you have just one weight-bearing leg you are forced to move the paired outer 'legs' (the crutches) first, followed by the one in the centre (your real leg). The only latitude is in whether you move your leg just as far forward as the crutches, or past them.

Walking with crutches uses up energy at a rate that is typically closer to that of running than walking, indicating that the use of crutches is not an especially energy-efficient way of getting about. Of course, unlike real legs, crutches do not have joints and elastic tissues that can store and release energy to optimise their efficiency, so the potential to evolve an efficient gait using three legs may well exist.

David Gillo
Chatham, Kent, UK

Kangaroos have strong tails capable of bearing weight, and though they do not have any 'three-legged' gaits, they can move slowly with a 'five-legged' gait. First the tail and forelegs are used to support the animal while the hind legs are brought forward in unison, then the hind legs take the weight while the kangaroo shifts forward before putting its forelegs and tail back onto the ground. Because the forelegs are short, the head stays close to the ground throughout, making this gait good for grazing.

The first vertebrates to walk evolved from fish, which swim with a lateral motion, so the gait they evolved probably also involved side-to-side movement. If fish had evolved differently, swimming with a vertical tail motion like a dolphin, then the first vertebrates would have had a gait with some up-and-down motion, possibly using the tail as a 'leg'. In this alternate reality, a five-legged gait similar to a grazing kangaroo could have been common, and tripedal creatures could conceivably have evolved.

Stuart Henderson
Farrer, Australian Capital Territory, Australia

Why are there only two sexes?

In a clumsy effort to seduce her, I was trying to explain the evolutionary advantages of sexual reproduction to a female friend the other day, one of which, I said, was introducing an element of genetic competition into the process. She wanted to know why, if two sexes are needed to create genetic competition, there aren't three, four or a million sexes to create even more competition. Why are there only two?

Tim Rowland
Bristol, UK

Some species do have more than two types. Single-celled ciliates have up to 100, and mushrooms have tens of thousands. But most organisms – even single-celled ones – come in two types.

So why are there two types in most species? In all species, no matter how many types, sex occurs between just two cells and any can mate with any other sex cell that is different from it. So, as your questioner suggests, finding just two types in most species is paradoxical, because having many types would maximise the chances of finding a mate.

One answer to this problem is that two types is best for coordinating the inheritance of cytoplasmic DNA – the part of the cell's genetic material that is not contained in the nucleus. However, there is a drawback to this solution. The species with two types fuse cells and potentially run the risk of scrambling this extra material.

The species with more than two mating types do it differently. With three types, the coordination is even more difficult to make error-proof, while those with many mating types don't fuse cells at all and so are not constrained to having just two types.

Laurence Hurst
Professor of Evolutionary Genetics
University of Bath, Somerset, UK

Having taught a difficult lesson on statistical techniques in geography to my secondary school students, I stood before them lost in admiration of the chi square test I'd written up on the board. Just then my students informed me that there were actually three sexes in this world: men, women and geography teachers. Unfortunately, I am a geography teacher.

Mary Sinclair
By email, no address supplied

Iron deficiency is common among human vegetarians, so how do herbivores cope?

Melanie Green
Hemel Hempstead, Hertfordshire, UK

Vegetarians have dietary difficulties because they force their omnivorous physiology to cope with a herbivorous diet, mineral imbalances being only one of the consequences.

Herbivores survive in good health partly because some are not as vegan as we might imagine. They eagerly eat animal dung, old bones, incidental insects and the like. They are also not too proud to eat dirt wherever they find a salt lick. Also, practically all herbivores rely on a partnership with gut flora to supply micronutrients or improve digestion.

Then again, they need to eat huge volumes of vegetation to ensure they absorb sufficient quantities of minerals from the minute concentrations in plants. After all, plants contain a little iron and manganese as well as macronutrients such as magnesium because these are needed for photosynthesis.

Humans trying to match the performance of specialist herbivores would need bellies like proboscis monkeys, and would be eating eighteen hours a day just to keep up; never mind the consequent activity at the nether end, nor the tooth wear that, as brachydont herbivores, humans would suffer.

Jon Richfield
Somerset West, Western Cape, South Africa

Do foxes and magpies communicate in a secret code?

From a slow-moving train I saw a fox standing with its tail resting on the ground while two magpies repeatedly took turns to peck the tip of the tail, before running off. The fox merely flicked its tail each time. What were they all doing?

Sue Murdochs
Marton cum Grafton, North Yorkshire, UK

Although counter-intuitive, the magpies that you observed may have been trying to prevent themselves becoming the fox's next meal. Animals have evolved a variety of ways to avoid predation. One strategy used by a number of species is to advertise that you are so fit and healthy that it would be pointless for the predator to waste its time trying to catch you. The magpies may have been doing just that.

Alternatively, if you saw this behaviour in the breeding season, the magpies may have been trying to protect their young. Many bird species will do everything possible to prevent predators from approaching their nesting sites or their newly fledged chicks. Birds achieve this either by acting as bait in order to lure the predators away from their offspring, or by directly attacking predators to drive them away. The ferocity with which birds attack predators often increases as the predator approaches the nest.

Magpies can become very aggressive towards intruders, so it is likely that the fox that you saw wasn't close enough to be a real threat, but was being given a friendly warning.

John Skelhorn
The Institute of Neuroscience
Newcastle University, UK

I have a magpie nest in my garden. Every spring, when the chicks hatch, both parents harass my two cats to prevent them approaching the tree. Sometimes the cats can hardly get out of the house. These magpies are quite fearless, chasing the cats right to my doorstep, and they keep up a loud and disagreeable chatter.

This behaviour lasts about a month, but can also happen in autumn or winter, even though the nest is empty. The magpies seem only to attack potential predators in pairs, but other members of the crow family are known to gang up against predators. I've seen flocks of crows attacking a buzzard or an owl.

Annelise Roman
By email, no address supplied

I have rescued sick and orphaned wildlife for thirty-nine years, specialising in rooks for the past decade, and have raised and released more than 200. Magpies and rooks are members of the crow family, and I have often seen the behaviour mentioned.

My dog is tolerant enough to let juvenile rooks ride on his back, and they will tease him into interacting with them. Some of the really mischievous characters among them play exactly the same game observed by the questioner. When my dog is napping they will tug at the end of his tail or ears, and jump out of reach when he reacts by flicking his tail or by shaking his head.

Magpies take up to three years to mature, and I have five non-flying resident adults to act as tutors. Without them I would be releasing a crowd of juvenile delinquents back into the wild.

The game of tail-tweaking described above was possibly being performed by young magpies and, if they are like my rook babies, they were probably doing it for fun. Among their many games, this is a favourite.

Tina Kirk
Swaffham, Norfolk, UK

How long would it take an average cow to fill the Grand Canyon with milk?

Nicola Stanley
Cambridge, UK

Obviously the first job would be to divert the Colorado river, which would otherwise interfere with the process. Secondly, the canyon would need to be dammed to retain the milk. Thirdly, because this is a desert environment, huge refrigeration capacity will be required to prevent the milk turning to cheese. And finally, to prevent loss of liquid by evaporation, the canyon will need to be hermetically sealed.

So, preparation complete, let's wheel in Daisy, the average cow. In the UK average milk yield per day per cow is in the range 15–20 litres. So let's settle on 17.5 litres. The canyon is 446 kilometres long by an average of sixteen kilometres wide and 1.6 kilometres deep, which gives a volume of about 10 million billion (1016) litres. So by simple division Daisy would take about 1.8 million million (1.8 × 1012) years to fill the canyon. This assumes the canyon has a rectangular cross section; for a triangular cross section, the time would be halved.

Now, suppose you don't want to wait 300 times the age of the planet for your canyon full of milk. Instead, you could divert the world's entire milk production to the canyon. This adds another requirement – a milk pumping infrastructure of epic proportions – unless you choose to use dried milk, which would be cheaper to transport, and then rehydrate it with water from the river. The UN Food and Agriculture Organization estimates that global milk production in 2004 was 504 million tonnes, which is equivalent to 489 billion litres, giving an estimated fill time of only about 20,000 years – still a pretty long job.

Jon White
Rampton, Cambridgeshire, UK

It all depends upon the size of the tanker truck the cow chooses to drive, the time it would take to drive from the milk distribution point, the inflow and outflow of the tanker truck, the ability to change the absorption and evaporation rates of the milk, and the ability of said cow to effectively block the exit route of the Colorado river.

Other considerations, of course, would be whether the cow works an eight-hour day or 24/7, and whether she ever has a day off. In a tangential vein, what subsidies would the US government be giving to the dairy farmers? This could be the making of another watershed in reality TV.

Bob Friedhoffer
New York City, USA

Do elephants sneeze?

Robin Rhind
London, UK

I frequently camp in the bush close to where I live in northern Botswana. By far the most pleasant way to experience the African night is to sleep under a mosquito net rather than in a tent, though you may end up, as I have, being investigated by lions, hyenas, hippos and elephants, which can be quite exciting.

A friend told me of a time when, sleeping under a net, he woke in the middle of the night and, not being able to see the stars, believed that it had clouded over and might rain. But as his eyes focused more clearly he realised that he was looking up at the underside of an elephant. Being inquisitive, the elephant was sniffing him through the net. Then, suddenly, there was an eruption from the elephant's trunk and my friend's face was covered in elephant mucus! The animal then carefully stepped over the net and went on its way.

Sneezing is an involuntary response that serves to remove foreign or excess material from the nasal passages. Elephants are just as liable to experience foreign matter in their nasal passages as other mammals and presumably sneeze for the same reason as dogs, cats and humans.

So yes, elephants do sneeze.

John Walters
Rakops, Botswana

I have experienced this at the zoo, feeding the elephants. Some children were also feeding one of them and, instead of holding out a handful of food, one of them held out a handful of pepper.

The elephant reached out, vacuumed the pepper right up its trunk, and what happened next was not pretty. The elephant snorted and wheezed a few times, and then flew into a sneezing 'rage' that resembled hurricanes being forced through a flailing fire hose.

The kids were thrown out of the zoo and, thankfully, the elephant was fine once its epic sneezing attack had concluded.

By email, no name or address supplied

Elephants do indeed sneeze. The thick lining of their trunks makes it more difficult for a substance to irritate the membranes inside, but some African farmers have discovered that chillies are an ideal elephant deterrent. In an effort to protect farmers' crops from elephants, which would otherwise damage or eat their produce, chilli seeds are planted around crop fields and dung cakes laced with chilli are burned at night.

A charity called the Elephant Pepper Development Trust actually helps develop the use of chilli and chilli oil in order to induce sneezing in elephants.

Georgia
By email, no address supplied

Do animals get travel sick?

On a long motorway journey while driving behind a horsebox, I wondered, do horses get travel sick? In fact, do we know whether any animals besides humans suffer from motion sickness?

Neil Bowley
Newthorpe, Nottinghamshire, UK

Horses are unable to vomit, except in extreme circumstances, because of a tight muscle valve around the oesophagus. So it is difficult to know whether or not they feel sick. Other monogastric animals can vomit. Younger cats and dogs frequently vomit during their first car journeys but rapidly become accustomed to travel and no longer suffer sickness. In the UK, a neurokinin-1 receptor antagonist has recently been licensed as a treatment for motion sickness in dogs as it reduces the urge to vomit.

James Hunt
Taunton, Somerset, UK

Motion sickness is common among animals, affecting domestic animals of all kinds. A carsick dog is not only pathetic, but messy. In his unforgettable book, *A Sailor's Life*, Jan De Hartog wrote: 'My worst memories of life at sea have to do with cattle. Two things no sailor will ever forget after such an experience are the pity and the smell . . . cattle get seasick, and the rolling of the ship terrifies the wits out of them. A seasick monkey or pup may be amusing and easy to deal with, but five hundred head of cattle in the throes of seasickness are a nightmare . . .' He also mentioned horses explicitly. Even fish transported in unsuitable conditions may show signs of disorientation.

Motion sickness is ubiquitous because all vertebrates have organs of balance and they correlate balance with feedback from other senses to stay upright. When movement causes say, visual information to conflict with balance, the brain of a sensitive individual interprets the disorientation as a symptom of poisoning and a typical reaction is to vomit to clear the gut.

Robert Amundsson
Copenhagen, Denmark

Both Robert Falcon Scott and Ernest Shackleton took ponies with them to Antarctica. On the way they experienced some appalling weather, and both noted how badly affected their animals were. They did, however, perk up when the storms abated. Similarly, Scott's dogs spent most of the storms curled up or howling, suggesting they, too, were suffering. Animals with a similar auditory system to ours would suffer from motion sickness, because it is caused by the confusion of auditory and visual inputs.

Tim Brignall
Bristol, UK

Does the animal kingdom have doctors?

I know that some animals treat simple injuries by licking them. Are there any animals that, like humans, treat each other's injuries, and do any animals have more sophisticated forms of 'medical treatment'?

David Taub
Karlstad, Värmland, Sweden

Licking each other's and their own wounds is the most common form of wound treatment for mammals. It is believed that such behaviour dates from the earliest days of mammals.

Saliva generally is germicidal and benefits wound tissue, causing little harm to live tissue while helping to slough off or recycle dead tissue.

The habit no doubt developed out of a defensive response to the pain, plus an eating response to bodily fluids and detritus. In fact, when mothers of many species lick sick cubs, if there is no improvement, it can lead to them eating their babies. Distressingly, such disruption may also lead the mother to eat the rest of the litter, especially if they are very young.

Formal hygiene and treatment of illness and injury, especially of other individuals, is mainly a human behaviour. However, it depends on what you choose to call 'treatment'. Candidate activities among birds include dust-bathing, hiding and resting when ill, and 'anting' – where they rub their feathers with ants, which then secrete antimicrobial chemicals. Various birds and mammals eat clays to counteract poisons in food, and some types of chimpanzee chew certain pungent leaves when ill. Such 'medicines' may control parasitic worms. Since plants and traditions vary by region, those habits clearly get passed on as learned knowledge.

Jon Richfield

Somerset West, Western Cape, South Africa

Saliva contains a complex cocktail of enzymes, many of which have antibacterial properties. In addition, it contains 142 epithetlial growth factors that promote healing in the wound; and the act of licking will tend to debride and remove gross contamination from the affected area. At the same time, of course, saliva contains huge numbers of various bacteria. Fortunately these are largely beneficial or have no effect, and there is no evidence to suggest they are detrimental to wound healing.

D. L. Harris
By email, no address supplied

Peruvian macaws are known to eat clay from riverbanks in behaviour known as geophagy. Animals and birds are often observed behaving in this way, but this is usually to provide grits for grinding food in their gizzards or supplying essential minerals to their diet. The macaws, however, consume only one particular type of clay, which is low in biologically relevant minerals and also far too fine to exert crushing and grinding effects on food in the gizzard. Instead, the birds are self-medicating to protect themselves from poisons. The clay is positively charged, and in the birds' stomachs it binds to negatively charged toxic alkaloids that have been ingested from unripe fruit and seeds. This protects the parrots from the effects of the alkaloids, while giving the macaws an ecological advantage over other animals and birds which cannot consume the same unripe foods.

Patrick Walter
London, UK

A previous correspondent wrote 'saliva contains huge numbers of various bacteria. Fortunately these are largely beneficial or have no effect, and there is no evidence to suggest they are detrimental to wound healing.'

On the contrary, there is compelling evidence to suggest that these bacteria (including *Streptococcus* and *Pasteurella* species) can colonise wounds and severely compromise healing. The use of Elizabethan-collars in dogs and cats to protect both surgical and traumatic wounds is as a consequence of this detrimental effect, and I would strongly discourage pet owners from allowing their pets to lick their wounds.

An alternative but more likely explanation for the behaviour described in the question is that carnivorous animals enjoy the act of licking a wound for the same reason that they enjoy the act of licking a bone: they like the taste.

Mike Farrell
European Specialist in Small Animal Surgery
University of Glasgow, UK

Do polar bears get lonely?

I'm not being flippant, just attempting to find out why animals such as humans or penguins are gregarious while others, such as polar bears and eagles, live more solitary lives.

Frank Anders
Amsterdam, The Netherlands

Having a gregarious or solitary nature are species-specific survival strategies adopted by different animals and birds.

Big predatory mammals such as polar bears, grizzlies and tigers isolate themselves from one another to avoid competition with other members of their own species. By spreading out, they also expand their feeding grounds and breeding territories. If fellow species members come into close proximity there can be fierce competition for food, mates and territory. The same is true with many solitary species of bird, such as eagles and condors.

These animals and birds usually pair up during the breeding season to reproduce, and separate soon after successful mating or when they have raised their young ones. In most cases, raising the young is the sole responsibility of females. Indeed, males of such species sometimes kill their young to increase their own reproductive success.

Social animals, by contrast, find strength in numbers. Animals such as antelope on the African savannah or penguins in the Antarctic form big colonies, where they huddle together for warmth and to alert each other to a potential predator attack. In a large herd or colony, losses to predators are negligible compared with what they would be if the animals were in isolated groups.

Between the solitary and social extremes are creatures like lions, wild dogs and wolves, which often hunt in groups and display differing degrees of social interaction and cooperation.

A similar question can be asked about why some plants are gregarious while others are solitary. In one intriguing strategy, called allelopathy, gregarious plants secrete chemicals into the soil to reduce competition from related species that cannot survive the presence of these compounds.

As with animals, these strategies have evolved to maximise the plants' chances of survival.

Saikat Basu
Lethbridge, Alberta, Canada

Bears and eagles rarely associate with their own kind because individuals need to defend their own feeding territories, in which food is often scarce. Polar bears live in an environment where the food resources are too limited to sustain a large community, so it makes sense for them to be the only predator in this particular niche. When food is plentiful both bears and eagles will gather together with a reasonable degree of amity.

The reverse is true for social animals, including humans. Social animals are often prey for other species, and cluster together for safety against predation – though this is only one of the reasons for group formation. But when food is scarce, individuals may break away from the group to find it.

Whether an animal can feel anything resembling the loneliness humans feel is hard to say. However, highly social animals, such as certain types of parrot, seem to be adversely affected when kept alone. Some parrots will engage in bizarre behaviours and can self-mutilate. Some large parrots will even seem to go 'insane' if subjected to long periods of isolation.

On the other hand, certain animals that are by nature solitary hardly appear to be affected at all. Some fish, in particular some types of cichlids, will fight viciously with their own kind if more than one is kept in an aquarium.

Guam rails, a kind of flightless bird, are notoriously

intolerant of their own kind, which has obviously made breeding them in captivity very difficult.

So, the answer to the question is a qualified yes: some animals will feel 'lonely' if they are by nature highly social. However, some will only engage with their own at specific times and in a highly ritualistic fashion, such as when mating or defending their territory.

By email, no name or address supplied

It depends on the bear and the circumstances. Loneliness is a reaction to deprivation of company when company is appropriate. In the case of polar bears, company usually represents competition or threat, so they do very well by themselves, thank you – unless you happen to be small enough to eat. In certain situations, when food and breeding are not relevant, males will wrestle harmlessly to establish dominance, thereby reducing the risks of dangerous fighting when mating time comes, but that is pretty much that.

Cubs want their mother's company for food, protection and reassurance, and they want each other's company for socialisation, warmth and play. Females want the company of their cubs, but keep other adults (and cubs) at a distance. Once her cubs mature or die, a mother again becomes a loner until mating time, and then tolerates males only briefly. She has no reason to want any company beyond that.

It is all part of the adaptation to their environment. In zoos, where security and food are no constraint, polar bears often seem happy to have the stimulus of company.

Jon Richfield

Somerset West, Western Cape, South Africa

Creature Curiosities

Animal pleasure

In 1995, primate pornography was the entertainment on offer for bored gorillas at Longleat House near Warminster, UK. Samba and Nico romped over a quarter of a hectare of private island during the day and enjoyed all the comforts of home at night.

Their house was equipped with satellite TV, to which they were glued throughout the long winter evenings. Longleat's press officer Claire Keener told *New Scientist* that Samba and Nico preferred jungle movies and became especially excited when gorillas appeared on the screen. At the consenting age of thirty years old, both were allowed to watch blue movies of primates mating in the wild, in the hope that they would be aroused into mating.

Not a bad life: first-class room service, a wholesome and varied diet of fruit, vegetables and beech leaves, and a visit from a personal vet once a week. Lights out at bedtime ensured a comfortable night's sleep but despite the movies, there was no monkeying around.

Exhibitionist spiny anteater reveals bizarre penis

In 2007, the bizarre sex life of the spiny anteater was exposed when it was discovered that the male ejaculates using only one half of its penis. These findings about the creature's sex life may seem salacious but they could help shed light on an evolutionary mystery.

It seems that the way the mammal ejaculates is similar to the way reptiles do – by shutting down one side of its penis before secreting semen from the other side. Reptiles have a pair of male members called hemipenes for sex, and they use only one of the two during each act of copulation.

The spiny anteater (*Tachyglossus aculeatus*), also known as the short-beaked echidna, is a primitive mammal found in Australia and New Guinea. Like the platypus, it is a monotreme, laying eggs instead of bearing live young.

Monotremes have many features in common with reptiles, and the hope is that by studying them, scientists may find clues as to how mammals evolved. The spiny anteater, however, is notoriously difficult to observe in the wild and shows little enthusiasm for breeding in captivity, so, prior to 2007, nobody had managed to observe them ejaculate.

Fortunately, Steve Johnston of the University of Queensland in Gatton, Australia, and his colleagues inherited a male spiny anteater that was not so shy. The creature had been 'retired' from a zoo because it produced an erection when being handled at public viewing sessions, bemusing its visitors. By filming this animal, the researchers were able to describe the unique spiny anteater erection and ejaculation behaviour for the first time.

The spiny anteater's four-headed phallus had been puzzling

scientists. 'When we tried to collect semen by electrically stimulated ejaculation before, not only did we not get a single drop, but the whole penis swelled up to a four-headed monster that wouldn't fit the female reproductive tract, which has only two branches,' said Johnston. 'Now we know that during a normal erection, two heads get shut down and the other two fit.' The heads used are swapped each time the mammal has sex.

The evolutionary significance of one-sided ejaculation was unknown, but may play a role in sperm competition – where sperm from many males may compete to fertilise an egg. Indeed, in the spiny anteater, up to eleven males may form a queue behind one female to copulate with her. The researchers also observed that hundreds of sperm team up to form bundles that swim much faster than individual sperm in the spiny anteater's semen – another possible adaptation for sperm competition.

'We can now study echidna sperm much better, which should offer fascinating insights into the evolution of mammals,' said Russell Jones from the University of Newcastle in New South Wales, who first dissected sperm bundles from dead echidna in the 1980s.

Bleeping Miss Daisy

In 1993, according to *New Internationalist*, cowherds in Japan were in short supply, so researchers tried developing pagers to attach to cows' collars. They could then be 'bleeped' at milking time. The cows responded to musical notes, and were apparently particularly attracted by piano melodies. Only two weeks were needed to train the cattle to come home when paged.

For farmers who were unwilling to invest in this technology, there was a much cheaper method – simply playing the appropriate piano tunes over loudspeakers at milking time. And to keep on playing until the cows came home.

Germany bans 'rejuvenating' sheep cell injections

In 1997, Germans were no longer permitted to have tissue from sheep foetuses injected into their buttocks. Several thousand people in Germany underwent this process every year, believing that the foetal cells had rejuvenating properties. But in the late 1990s, health officials in Bonn said that they had decided to ban the practice, pointing to severe immune reactions in some patients. A Ministry of Health report said that up to 5 per cent of patients had reactions to the injections and in five documented cases patients had died.

Jutta Buscha, a doctor in the Bavarian town of Rottach-Egern who practised the therapy, said, 'I don't understand this enmity, this persecution. The people who sit in judgment on us never even came for a visit.'

INDEX